ENVIRONMENTAL NOISE CONTROL

ENVIRONMENTAL NOISE CONTROL

EDWARD B. MAGRAB, PH.D.

Department of Civil and Mechanical Engineering
The Catholic University of America
Washington, D. C.

A Wiley-Interscience Publication

JOHN WILEY & SONS
New York London Sydney Toronto

Library of Congress Cataloging in Publication Data:

Magrab, Edward B.
    Environmental noise control

    "A Wiley-Interscience publication."
    Includes bibliographical references and index.
    1. Noise pollution. 2. Noise control. I. Title.

TD892.M33   620.2′3   75-20233
ISBN 0-471-56344-7

Printed in the United States of America

10 9 8 7 6 5 4 3 2 1

PREFACE

This book presents a modestly comprehensive synthesis of the large amount of information pertaining to noise control that has become available in recent years. The approach is a compromise between an introduction to the subject and the detailed and exhaustive presentations found in handbooks. The book itself is an outgrowth of lecture notes prepared for a senior-level/first-year-graduate-level course entitled "Environmental Noise and Its Control" taught from 1970 to 1974 in the School of Engineering and Architecture at The Catholic University of America.

The book is reasonably self-contained, with some prior knowledge of elementary algebra and the logarithm assumed. Virtually all the results are presented either in simple algebraic form or graphically. Since no derivations are given, care has been taken to indicate the limitations of the formulas and, where appropriate, to discuss their physical significance. In addition, numerous example problems are presented to illustrate virtually all the important results. The first chapter presents a brief summary of a few results from theoretical acoustics and introduces several definitions and physical quantities required in the subsequent chapters. Chapter Two provides a summary of the current data indicating the risk of hearing damage from noise as a function of its level and spectral content. The third chapter describes numerous single number metrics that attempt to describe the sociological and psychological responses to various types of noises. Chapter Four is a brief but comprehensive presentation of the fundamentals of electronic instrumentation, and includes a detailed discussion of microphones and sound-level meters. The fifth chapter contains an extensive compilation of empirical results from which an estimate of the sound-pressure or sound-power levels of common noise sources can be obtained. Chapter Six defines the various acoustic absorption coefficients and describes how absorption can be used to change the overall acoustics of an enclosure. The last chapter presents four general methods of noise and vibration control: mufflers, barriers, vibration isolators, and the transmission loss and impact isolation of panels.

v

I wish to express my appreciation and thanks to the following persons for their extremely helpful comments concerning the final manuscript: Raymond D. Berendt, Donald S. Blomquist, Curtis I. Holmer, Dr. John A. Molino, and Dr. David S. Pallett. My thanks and deep appreciation are also extended to Dr. John J. Gilheany, who provided invaluable assistance in the formative stages of the manuscript. I am also indebted to Liisa Srholez, Cynthia Pilot, and Carolyn Smith for typing the various drafts of the manuscript and to Jafar Vossoughi for the art work.

EDWARD B. MAGRAB

Washington, D. C.
September 1975

CONTENTS

1

INTRODUCTION

1.1 INTRODUCTION

This chapter covers the elementary and basic concepts of acoustics. It introduces the fundamental properties of propagating plane and spherical waves, the physical quantities of interest (sound pressure and sound power) and their units, and the definition and manipulation of the decibel.

1.2 BASIC PROPERTIES OF WAVES

Introduction

Gases, such as air, have both mass density and volume elasticity. The elasticity causes the gas to resist being compressed, tending to return itself to its original state upon release of the compressing forces. The inertia of the mass density causes the motion of the gas to "overshoot" its original equilibrium position upon release of these forces. Both of these properties, mass density and volume elasticity, are the two requisites of wave motion and, therefore, necessary for a medium to propagate a sound wave. At equilibrium the gas has a density (kg/m^3), is under a uniform pressure (pascals), and is at a uniform temperature ($^\circ K$). These three quantities are related by an equation of state. It can be shown that a sound wave produces changes in each of these three quantities, each change being proportional to the amplitude of the sound wave. Of these three quantities the easiest one to measure most accurately is the change in pressure.

Plane Waves

Consider a sound wave of frequency f (measured in Hz) which propagates outwards from a source such that no waves are reflected back to the source or in any way altered from its original outward direction. In addition it is assumed that as the wave propagates, its amplitude remains constant. Let the source be idealized as a rigid piston in a very long tube. As the piston moves forward from the equilibrium position, the air in front is compressed. This creates a wave whose pressure is greater than that of the atmosphere. As the piston moves backwards past the equilibrium position, the air in front of the piston moves backwards to fill the increased volume caused by the piston's displacement. Since the volume is now larger, the pressure is now less than atmospheric pressure and, consequently, the air has been rarefied. This process repeats itself every period, T, and is related to the frequency by

$$T = \frac{1}{f} \qquad \text{sec} \tag{1-1}$$

(The frequency, f, is related to the angular or radian frequency, ω, by $\omega = 2\pi f$ rad/sec.) The mean portion of this infinitesimally small volume of air molecules remains stationary in space and only the compression and rarefaction propagate from the source at a characteristic speed of the medium, c, called the speed of sound.

Since the sound is traveling in space, it too has a spatial equivalent to the period. It is called the wavelength of sound, λ, and is given by

$$\lambda = \frac{c}{f} = cT = \frac{2\pi c}{\omega} = \frac{2\pi}{k} \qquad \text{m} \tag{1-2}$$

where we have introduced another quantity, k, called the wave number. The wavelength has the dimensions of length and the wave number the dimensions of the reciprocal of length. This concept of wavelength is fundamental to understanding the propagation and attenuation of sound in general. The various ideas described above are depicted graphically in Figure 1-1.

The speed of sound in air is a function of temperature. If it is assumed that air behaves as an ideal gas, which for all practical purposes it does, the equation for the speed of sound in air is given by

$$c = 20.05 \sqrt{T_0} \qquad \text{m/sec} \tag{1-3}$$

where T_0 is the absolute temperature in degrees Kelvin ($273.2° + °C$). At a temperature of $21°C$ we find from (1-3) that $c = 344$ m/sec. Table 1-1 gives

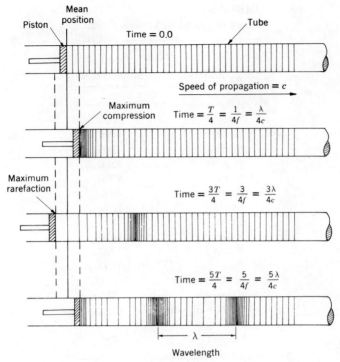

FIGURE 1-1
Propagation of sound waves in a long tube caused by a piston oscillating at a frequency f. (Reproduced from *Building Acoustics* by Day, Ford, and Lord by courtesy of the publisher and authors.)

values of the speed of sound in various materials and gases. Table 1-2 gives ω, λ, and k as a function of selected frequencies for sound propagation in air at 21°C.

Spherical Waves

Consider a sphere of radius b whose surface is vibrating with a uniform radial expansion and contraction at a frequency f. If the source is radiating into a medium such that no sound waves are reflected back in the direction of the source, and if the product of kb, where k is the wave number in the medium surrounding the sphere, is much less than unity, then the sound pressure at any radial distance r from the sphere is inversely proportional to r. That is

$$p \sim \frac{1}{r} \tag{1-4}$$

TABLE 1-1
Speed of Sound in Various Media

Material	Speed of Sound (m/s)
Aluminum	5820
Brick	3600
Concrete	3700
Steel	4905
Copper	4500
Glass	5000
Iron	4800
Lead	1260
Tin	4900
Water	1410
Wood	3300
Zinc	3750
In some gases	
Ammonia	415
Carbon dioxide	258
Hydrogen	1270
Steam (100°C)	405

In other words, each time the distance from this type of sound source is doubled, the sound pressure is halved. As shown in Section 1.5, this corresponds to a 6 dB decrease in the sound pressure level for each doubling of distance. A sound source for which $kb \ll 1$ and (1-4) holds is called a *point source*. A point source radiates uniformly in all directions.

In a certain range of values of r from a source the sound pressure and the particle velocity will be in phase. When this occurs, and (1-4) is valid, r denotes a position in the *far-field*. In the far-field only a single measurement is required to determine all the characteristics of the sound field. The *near-field* is that region wherein the particle velocity and pressure are not in phase. In practice most noise sources cannot be classified as simple point sources. However, the sound field of a complicated sound source will look as if it were a point source if the following two conditions are met: (1) $r/b \gg 1$, that is, the distance from the source is large compared to its characteristic dimension, and (2) $b/\lambda \ll r/b$, that is, the ratio of the size of the source to the wavelength of sound in the medium is small compared to the ratio of the distance from the source to its characteristic dimension.

TABLE 1-2
Radian Frequency, Wavelength, and Wave Number at Selected Frequencies[a] for c = 344 m/sec

f(Hz)	ω(rad/sec)	λ(m)	k(m⁻¹)
25	157	13.76	0.456
31.5	197	10.92	0.575
40	251	8.60	0.730
50	314	6.88	0.912
63	395	5.46	1.150
80	502	4.30	1.460
100	628	3.44	1.825
125	785	2.75	2.283
160	1,004	2.15	2.920
200	1,256	1.72	3.651
250	1,570	1.37	4.56
315	1,970	1.09	5.75
400	2,510	0.86	7.30
500	3,140	0.69	9.12
630	3,950	0.55	11.50
800	5,020	0.43	14.60
1,000	6,280	0.34	18.25
1,250	7,850	0.27	22.83
1,600	10,040	0.22	29.20
2,000	12,560	0.17	36.51
2,500	15,700	0.14	45.6
3,150	19,700	0.11	57.5
4,000	25,100	0.08	73.0
5,000	31,400	0.07	91.2
6,300	39,500	0.06	115.0
8,000	50,200	0.04	146.0
10,000	62,800	0.03	182.5
12,500	78,500	0.03	228.3
16,000	100,400	0.02	292.0
20,000	125,600	0.02	365.1

[a]These frequencies are the center frequencies of the standard 1/3-octave bands. See Section 1.8 and Table 1-3.

Recall that $r/b \gg 1$ from the first condition. A value of $r/b > 3$ is a sufficient approximation; therefore, $b/\lambda \ll 3$.

Effects of Atmospheric Conditions on Sound Propagation

As shown in the preceding section, under certain conditions the sound pressure level decreases 6 dB for each doubling of distance from the source. There are however several climatological conditions that alter this rate: air absorption, temperature gradients, and wind.

Sound absorption in quiet, isotropic air is caused by two processes. First, energy is extracted from a sound wave by losses arising from heat conduction and viscosity in the air. This type of absorption is significant only at very low temperatures. Second, energy is extracted from a sound wave by rotational and vibration relaxation of the oxygen molecules in the air. This molecular absorption depends in a major way on temperature. The dependence of the sound absorption on frequency, temperature, and humidity has been determined theoretically, in the laboratory, and in the field. The results, in one form, are the air-to-ground-attenuation charts standardized by the aircraft industry and shown in Figures 1-2 and 1-3. These figures clearly show that for a given percentage humidity high frequencies are attenuated more rapidly than low frequencies. Also, the maximum attenuation for any frequency occurs when the humidity is the least.

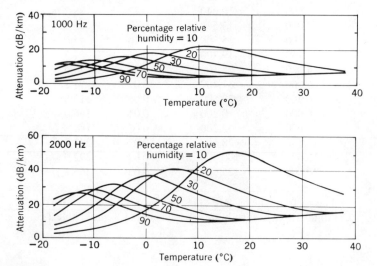

FIGURE 1-2

Atmospheric attenuation for air-to-ground propagation for octave bands with center frequencies of 1000 and 2000 Hz.

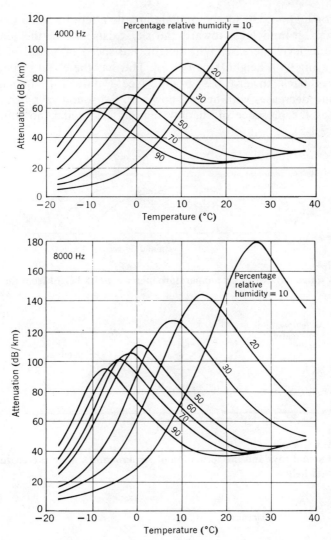

FIGURE 1-3
Atmospheric attenuation for air-to-ground propagation for octave bands with center frequencies of 4000 and 8000 Hz.

The propagation paths for sound are greatly altered when a temperature profile from ground level toward the sky exists. When the temperature from ground level increases with height, the speed of sound of the air is also increasing with height [recall (1-3)]. This has the effect of bending the sound rays back toward the ground (see Figure 1-4). When the air temperature decreases with height, however, the sound rays bend upward. This causes the presence of shadow regions into which no sound enters (see Figure 1-5).

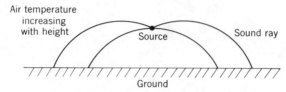

FIGURE 1-4
Bending of sound rays when air temperature increases with height from the ground.

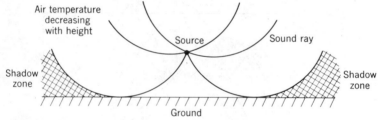

FIGURE 1-5
Bending of sound rays and the formation of shadow zones when the air temperature decreases with height.

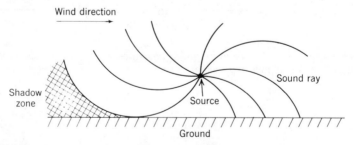

FIGURE 1-6
Bending of sound rays and the formation of a shadow zone because of wind.

The same reasoning may be applied to an atmosphere in the presence of wind, as shown in Figure 1-6. With a typical wind profile there will be a shadow region upwind and no shadow region downwind. This accounts for the oft-observed difficulty in hearing upwind from a source. It is important to emphasize that this is because the wind velocity normally *increases* with height.

An important difference between the effects of temperature and wind is that although reciprocity applies as far as the former is concerned, it does not hold for the latter. That is, a source and receiver at the ground may be interchanged and there will be no difference at the receiver when we have only a temperature gradient. But in the presence of a wind gradient the sound level at the receiver is much greater when it is in a downwind position than after the interchange of position.

1.3 RMS SOUND PRESSURE

The rms sound pressure is obtained by squaring the value of the amplitude of the sound pressure at each instant of time. These squared values are then summed and the total divided by the total time, T, over which these values were obtained. The rms value of the sound pressure, p_{rms}, is the square root of this time average. Since the original sound pressure is squared, the rms value is always non-negative. It can be shown that the square of p_{rms}, that is, p_{rms}^2, is proportional to the average power over the total sample time T. If the peak value of the sound pressure of a *single* tone is P_0, then $p_{rms} = P_0/\sqrt{2}$. This is the *only* time this relationship holds.

Consider the case of two pure tones where the amplitude of the first tone (of frequency f_1) is P_1 and that of the second tone (of f_2) is P_2 and $f_1 \neq f_2$. The rms value, $p_{T_{rms}}$ is

$$p_{T_{rms}} = \left[\left(\frac{P_1}{\sqrt{2}} \right)^2 + \left(\frac{P_2}{\sqrt{2}} \right)^2 \right]^{1/2} \qquad f_1 \neq f_2 \qquad (1\text{-}5)$$

Equation 1-5 can be generalized to include any number of tones, no two of which have the same frequency. Hence

$$p_{T_{rms}} = \frac{1}{\sqrt{2}} \left[\sum_{j=1}^{N} P_j^2 \right]^{1/2} \qquad f_j \neq f_i \qquad (1\text{-}6)$$

Notice that the rms measurement is independent, when $f_j \neq f_i$, of any phase difference between the signals. When two tones are of the same frequency we have

$$p_{T_{\text{rms}}} = \frac{1}{\sqrt{2}} \left(P_1^2 + P_2^2 + 2P_1 P_2 \cos \Theta \right)^{1/2} \qquad f_1 = f_2$$

where Θ is the phase difference between the two signals. When $\Theta = 0°$,

$$p_{T_{\text{rms}}} = \frac{(P_1 + P_2)}{\sqrt{2}}$$

is a maximum and when $\Theta = \pm 180°$ (exactly out of phase),

$$p_{T_{\text{rms}}} = \frac{(P_1 - P_2)}{\sqrt{2}}$$

is a minimum.

In the far-field the average intensity, that is, the average energy that flows through a unit area per unit time, is given as

$$I = \frac{p_{\text{rms}}^2}{\rho c} \qquad \text{W/m}^2 \tag{1-7}$$

where ρ is the density of the medium and c is the speed of sound in the medium. For air at 22°C and 0.750 m of Hg, $\rho c = 412$ N-sec/m^3.

1.4   POWER, INTENSITY, AND ENERGY DENSITY

A sound source radiates power, denoted by W. A portion of this power will flow through each small region of the medium that surrounds the source. If there are no losses in the medium, all of this radiated power must pass through any surface that encloses the source. The larger the enclosing surface, the less power per unit area that will pass through any element of the surface. The total sound power is sum of the products of the intensity, I_s, through each incremental area ΔS. The total power W is equal to $\Sigma I_s \Delta S$.

The intensity itself is difficult to measure. However, in the far-field the intensity is related to p_{rms} according to (1-7). Since we can measure p_{rms} easily, we often use (1-7) to obtain I.

The sound energy density is the energy stored in a small volume of air in an enclosure owing to the pressure of a standing wave field. (A standing wave is distinct from the freely progressing wave shown in Figure 1-1 in that the sound pressure at one place does not occur to the right or to the left of that place at the next instant. The wave no longer travels; it is a standing wave. To obtain a standing wave in the tube of Figure 1-1, the tube is terminated with a reflecting surface.) The relation between the space-average mean-square sound pressure p_{av} ($=$ average over space of p_{rms}) and the space-average sound-energy density, D, is

$$D = \frac{p_{av}^2}{\rho c^2} \qquad \text{W-sec}/\text{m}^3 \qquad (1\text{-}8)$$

This quantity D is used in room-acoustic calculations.

1.5   LEVELS AND THE DECIBEL

The purpose of the level scale is to transform the ratio of two power- (or intensity-) like quantities by taking the logarithm of this ratio. One of the quantities used to obtain this ratio is called the reference quantity. The argument of the logarithm is dimensionless, and the scale is said to give the level of the sound in dB above (or below) the reference level that is determined by the reference quantity. The denotation "dB" does not represent any physical unit, but merely indicates that a logarithmic transformation has been performed. It does, however, relate to physical units when followed by the statement "re U_0" which means referenced to the physical quantity U_0.

The sound-power level is defined as

$$L_w = 10 \log_{10}\left(\frac{W}{W_0}\right) \qquad \text{dB re } W_0 \qquad (1\text{-}9)$$

and, conversely,

$$W = W_0 \text{antilog}_{10}\left(\frac{L_w}{10}\right) = W_0 10^{(L_w/10)} \qquad (1\text{-}10)$$

where W is the sound power in watts and W_0 is the reference sound power in watts.

From the properties of logarithms we recall the following: If $A = B \times C$, then

$$\log_{10}(A) = \log_{10}(BC) = \log_{10} B + \log_{10} C \qquad (1\text{-}11)$$

If $A = B/C$, then

$$\log_{10}(A) = \log_{10}\left(\frac{B}{C}\right) = \log_{10} B - \log_{10} C \qquad (1\text{-}12)$$

Thus we see that adding decibels is the same as multiplication, and subtracting them is the same as division. It should be noted that if $A = 1$, then $\log_{10} A = 0$. This is easily deduced from (1-10) with $W_0 = W = 1$. Also, the logarithm of a number greater than one is always positive. If A is positive, but less than one, $\log_{10} A$ is a negative quantity. Using (1-12) we see that if, for example, $A = 0.5$, and, therefore, $B = 1$ and $C = 2$, then $\log_{10} 0.5 = -\log_{10} 2$.

The reference quantity appearing in (1-9) has, by international agreement, been established as $W_0 = 10^{-12}$ W. Hence (1-9) can be written as

$$L_w = 10 \log_{10}\left(\frac{W}{10^{-12}} \right) = 10 \log_{10} W + 120 \qquad \text{dB re } 10^{-12} \text{ W} \qquad (1\text{-}13)$$

The intensity level is defined as

$$L_I = 10 \log_{10}\left(\frac{I}{I_{\text{ref}}} \right) \qquad \text{dB re } I_{\text{ref}} \qquad (1\text{-}14)$$

where I_{ref} has been standardized as 10^{-12} W/m^2. Thus (1-14) can be rewritten as

$$L_I = 10 \log_{10} I + 120 \qquad \text{dB re } 10^{-12} \text{ W/m}^2 \qquad (1\text{-}15)$$

Most measuring instruments respond to sound pressure. In the far-field the square of the sound pressure is proportional (not equal) to sound power. Thus the sound-pressure squared level is usually shortened to

$$L_p = 10 \log_{10}\left(\frac{p_{\text{rms}}^2}{p_{\text{ref}}^2} \right) = 20 \log_{10}\left(\frac{p_{\text{rms}}}{p_{\text{ref}}} \right) \qquad \text{dB re } p_{\text{ref}} \qquad (1\text{-}16)$$

and, conversely,

$$p_{\text{rms}} = p_{\text{ref}} \, \text{antilog}_{10}\left(\frac{L_p}{20} \right) = p_{\text{ref}} \, 10^{(L_p/20)} \qquad (1\text{-}17)$$

where p_{rms} is the rms sound pressure in pascals (Pa) and p_{ref} is the rms reference sound pressure in pascals. The reference pressure, p_{ref}, is equal to 20 μPa $= 20$ μN/m^2 (micronewtons/m^2). Thus (1-16) becomes

$$L_p = 20 \log_{10}(p_{\text{rms}}) + 94 \qquad \text{dB re } 20 \ \mu\text{Pa} \qquad (1\text{-}18)$$

The quantity L_p is referred to as the sound-pressure level.

The numerical relationship between decibels and its argument expressed as a ratio can be simplified somewhat with the use of Figure 1-7. The left-hand ordinate corresponds to sound-pressure levels (L_p) and the right-hand ordinate to sound-power levels (L_w) or intensity levels (L_I). The abscissa gives the ratio, which is an ordinal number. For example, if the sound pressure level increases by a factor of 500 ($=5 \times 10^2$), then Figure 1-7 yields $14 + 40 = 54$ dB, since $N = 2$. On the other hand if the sound power level decreases by 17 dB then, since $-17 = 3 + 10(-2)$, the original quantity has been reduced by a factor of $2 \times 10^{-2} = 0.02$.

Example 1-1. A sound pressure is (a) doubled, (b) halved, (c) increased tenfold, and (d) decreased by a factor of ten. Express these changes as changes in sound-pressure levels.

Solution: Let $p_{rms} = p_1$ and $p_{ref} = p_2$. Then (1-16) can be written as

$$\Delta L_p = 20 \log_{10}\left(\frac{p_1}{p_2}\right) \quad \text{dB}$$

where ΔL_p denotes the change in dB.

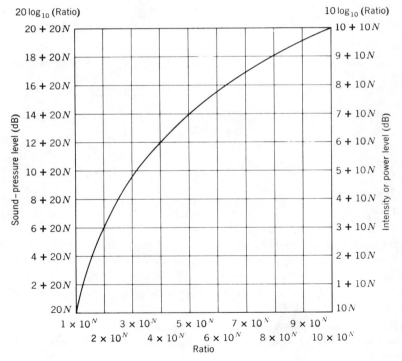

FIGURE 1-7
Graphical relationship between decibel and its ordinal counterpart.

(a) $p_1/p_2 = 2$

$$\Delta L_p = 20\log_{10}(2) = 20(0.3010) = 6.0\text{ dB}$$

(b) $p_1/p_2 = 1/2$

$$\Delta L_p = 20\log_{10}\left(\frac{1}{2}\right) = -20\log_{10}(2) = -6.0\text{ dB}$$

(c) $p_1/p_2 = 10$

$$\Delta L_p = 20\log_{10}(10) = 20(1) = 20\text{ dB}$$

(d) $p_1/p_2 = 0.1$

$$\Delta L_p = 20\log_{10}\left(\frac{1}{10}\right) = -20\log_{10}(10) = -20\text{ dB}$$

It should be noted from cases (a) and (b) that $+6$ dB corresponds to an increase of 100% above the original sound pressure whereas -6 dB corresponds to a decrease of 50% from the original sound pressure. These results can also be obtained from Figure 1-7.

Example 1-2. The sound-pressure level (a) increased 26 dB and (b) decreased 32 dB. Determine the change in the original sound pressure.

Solution: We shall determine the solution two ways. The first method uses (1-17). Thus

$$\text{(a)} \qquad \frac{p_1}{p_2} = 10^{(26/20)} = 10^{1.3} = 20$$

$$\text{(b)} \qquad \frac{p_1}{p_2} = 10^{(-32/20)} = 10^{-1.6} = 0.025$$

The second procedure uses the results of Example 1-1. For case (a) we note that 26 dB = 20 dB + 6 dB. A change of 20 dB corresponds to an increase by a factor of 10, and a change of 6 dB to an increase by a factor of 2. Thus a change of 26 dB corresponds to a twentyfold (10×2) increase in the original sound pressure. For case (b) we note that -32 dB = -20 dB -6 dB -6 dB. Since -20 dB corresponds to a decrease in the original sound pressure by a factor of 10, and each -6 dB to a halving of the original sound pressure, -32 dB corresponds to a sound pressure that is $1/40$ ($=1/2 \times 1/2 \times 1/10$) of its original value. A third method is to use Figure 1-7.

Example 1-3. A sound pressure is increased by 20%. Express this increase as a change in sound-pressure level.

Solution: We note that percent change$=[(p_2-p_1)/p_1]\times 100$. Thus an increase of 20% corresponds to the ratio $p_2/p_1=1.2$. From (1-16)

$$\Delta L_p = 20\log_{10}(1.2) = 20(0.0792) = 1.58 \text{ dB}$$

Example 1-4.   A sound-pressure level of 80 dB re 20 μPa is measured. What is the absolute level of the sound pressure?

Solution: From (1-17)

$$p_{\text{rms}} = 20[10^{(80/20)}] \ \mu\text{Pa}$$

$$p_{\text{rms}} = 2\times 10^5 \ \mu\text{Pa} = 0.2 \text{ Pa}$$

1.6   RELATIONS AMONG SOUND-POWER LEVELS, INTENSITY LEVELS, AND SOUND-PRESSURE LEVELS

Using (1-7) and (1-14) we have that

$$L_I = 10\log_{10}\frac{I}{I_{\text{ref}}} = 10\log_{10}\frac{p_{\text{rms}}^2}{\rho c I_{\text{ref}}}$$

$$= 10\log_{10}\frac{p_{\text{rms}}^2}{p_{\text{ref}}^2}\frac{p_{\text{ref}}^2}{\rho c I_{\text{ref}}}$$

$$L_I = L_p - C_1 \qquad \text{dB re } 10^{-12} \text{ W/m}^2 \qquad (1.19)$$

where

$$C_1 = 10\log_{10}\frac{I_{\text{ref}}\rho c}{p_{\text{ref}}^2} = 10\log_{10}\frac{\rho c}{400} \qquad \text{dB} \qquad (1\text{-}20)$$

and L_p is given by (1-16).

As stated at the end of Section 1.3, at standard conditions $\rho c = 412$ N-sec/m^3. At these conditions $C_1 = 0.13$ dB, an amount that is usually not significant. Thus for far-field noise measurements

$$L_I \approx L_p \qquad (1\text{-}21)$$

for freely progressing waves. When the temperature, barometric pressure, or both, differ greatly from these standard conditions C_1 is determined using Figure 1-8.

When the intensity is uniform over an area S, the sound power and intensity are related by $W = IS$. Since $W_0 = I_{\text{ref}}S_0$, where S_0 is a reference

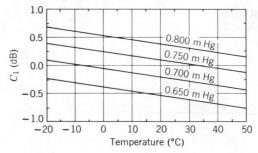

FIGURE 1-8
Value of C_1 as a function of temperature and barometric pressure.

area of $1\,m^2$, (1-9) becomes

$$L_w = 10\log_{10}\frac{W}{W_0} = 10\log_{10}\frac{IS}{I_{ref}S_0}$$

$$= 10\log_{10}\frac{I}{I_{ref}} + 10\log_{10}\frac{S}{S_0}$$

$$L_w = L_I + 10\log_{10}S \qquad \text{dB re } 10^{-12}\,\text{W} \qquad (1\text{-}22)$$

since $10\log_{10}S_0 = 10\log_{10}1 = 0$. If S also equals $1.0\,m^2$ then $L_w = L_I$.

Example 1-5   A sound source is found to be radiating uniformly over a spherical surface that is 2.0 m from its center. The recorded sound-pressure level at this distance is 78 dB re 20 μPa. The measurement was made at a temperature of 30°C and an ambient pressure of 0.700 m Hg. What is the sound power of the source?

Solution: From Figure 1-8 it is found that $C_1 = -0.3\,$dB. Using (1-19) in (1-22) yields

$$L_w = L_p - C_1 + 10\log_{10}S = 78 + 0.3 + 10\log_{10}16\pi$$

$$L_w = 95.3 \qquad \text{dB re } 10^{-12}\,\text{W}$$

where we have used the fact that $S = 4\pi r^2$ and r is the radius of the sphere.

1.7 MANIPULATIONS WITH DECIBELS

Sums and Differences

We are often required to estimate the total sound-pressure level of two or more noise sources when the sound-pressure levels of each source is known. If the sources are statistically unrelated noise sources, then we can use (1-5). Thus

$$\frac{p_T}{p_{\text{ref}}} = \left(\sum_{j=1}^{N} \frac{p_{j_{\text{rms}}}^2}{p_{\text{ref}}^2} \right)^{1/2}$$
(1-23)

or from (1-17)

$$10^{[L_T/20]} = \left(\sum_{j=1}^{N} 10^{(L_j/10)} \right)^{1/2}$$
(1-24)

where L_T is the total sound-pressure level and L_j is the sound-pressure level of the individual sources. Equation 1-24 is often tedious to use. However, if we have only two sources ($N = 2$) and we assume that $L_1 > L_2$, then (1-24) can be rewritten as

$$L_T = L_1 + 10\log_{10}(1 + 10^{-\{(L_1 - L_2)/10\}}) \qquad \text{dB re 20 } \mu\text{Pa} \qquad (L_1 \geqslant L_2)$$
(1-25)

Thus from (1-25) it is evident that the total sound-pressure level is the sum of the larger sound-pressure level (L_1) and a term that only depends on the numerical difference between the larger and smaller levels. Equation 1-25 is plotted in Figure 1-9.

Now consider the case wherein we are required to estimate the sound-pressure level of an unknown source when the total sound-pressure level and that of the second source are known. In this case (1-24) is written as

$$10^{(L_T/10)} - 10^{(L_1/10)} = 10^{(L_2/10)}$$

or

$$L_2 = L_T + 10\log_{10}(1 - 10^{-\{(L_T - L_1)/10\}}) \qquad \text{dB re 20 } \mu\text{Pa} \qquad (1\text{-}26)$$

where L_2 is the sound-pressure level of the unknown source. Since $L_T - L_1 > 0$ the argument of the logarithm is positive, but less than unity. As shown in Section 1.5, the logarithm of a number less than one is negative and, therefore, L_2 will be less than L_T, as it should be. Equation 1-26 is plotted in Figure 1-10.

Example 1-6   Three noise sources each generate a sound-pressure level of 83 dB re 20 μPa. What would be the total sound-pressure level caused by the three of them together?

Solution: Two methods can be employed to determine the solution. First, using (1-25) we have for two sources

$$L_T' = 83 + 10\log_{10}(1 + 1) = 83 + 10(0.3010)$$

$$L_T' = 86 \qquad \text{dB re 20 } \mu\text{Pa}$$

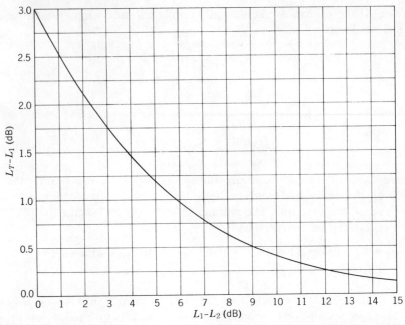

FIGURE 1-9
Chart for adding two uncorrelated levels L_1 and $L_2(L_1 > L_2)$. L_T is the total level.

Considering the above value the sound-pressure level of a new source we again employ (1-25) to obtain the total sound-pressure level for the three sources:

$$L_T = 86 + 10\log_{10}(1 + 10^{-[(86-83)/10]})$$

$$L_T = 86 + 1.76 = 87.76 \qquad \text{dB re } 20 \ \mu\text{Pa}$$

The total sound-pressure level could also be obtained directly from (1-24). Thus

$$L_T = 10\log_{10}(10^{8.3} + 10^{8.3} + 10^{8.3})$$

$$L_T = 87.77 \qquad \text{dB re } 20 \ \mu\text{Pa}$$

In the second method, the difference between two of the sources is 0 dB. Hence from Figure 1-9 we find that their combination will yield a sound-pressure level of 86 dB. We now consider this value the sound-pressure level of one source to which we "add" another source (the third source) with a sound-pressure level of 83 dB. Since 86 dB − 83 dB = 3 dB, Figure

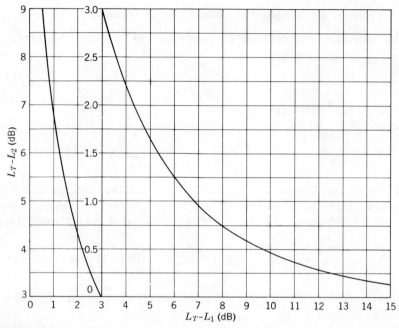

FIGURE 1-10
Chart for subtracting two uncorrelated levels L_T and $L_1(L_T > L_1)$. L_2 is the unknown level.

1-9 gives a total sound-pressure level for the three sources as 86 dB + 1.75 dB = 87.75 dB re 20 μPa.

Example 1-7.   An overall sound-pressure level of 90 dB re 20 μPa exists in a machinery room. When one machine is turned off the sound-pressure level drops to 85 dB. What is the sound-pressure level of the machine that was turned off?

Solution: Using (1-26) gives

$$L_2 = 90 + 10\log_{10}(1 - 10^{-\{(90-85)/10\}})$$

$$L_2 = 90 + 10\log_{10}(0.683) = 90 - 1.66 = 88.34 \qquad \text{dB re 20 } \mu\text{Pa}$$

In a second procedure the difference between the total sound-pressure level and the sound-pressure level with one machine turned off is 5 dB. Using Figure 1-10 we find the sound-pressure level of the machine turned off to be 90 dB − 1.66 dB = 88.34 dB. As a check we note that the sum of two noise sources, one having a sound-pressure level of 85 dB and the other 88.34 dB is, from Figure 1-9, equal to 90 dB.

Averages and Standard Deviations: Exact Formulas

The average value or mean of a collection of N sound pressures p_j, is*

$$\bar{p} = \frac{1}{N} \sum_{j=1}^{N} p_j \qquad (1\text{-}27)$$

which has a standard deviation of

$$\sigma = \frac{1}{\sqrt{N-1}} \left[\sum_{j=1}^{N} (p_j - \bar{p})^2 \right]^{1/2} = \frac{1}{\sqrt{N-1}} \left[\sum_{j=1}^{N} p_j^2 - N(\bar{p})^2 \right]^{1/2} \qquad (1\text{-}28)$$

In many applications in airborne acoustics the average value and standard deviation of numerous readings of the sound-pressure level is desired. Therefore, we make use of (1-16) and (1-17) in (1-27) to obtain

$$\bar{L}_p = 20 \log_{10} \frac{1}{N} \sum_{j=1}^{N} 10^{(L_j/20)} \qquad \text{dB re 20 } \mu\text{Pa} \qquad (1\text{-}29)$$

where $\bar{L}_p = 20 \log_{10} \bar{p}$ is the average sound-pressure level and L_j is the jth sound pressure level, $j = 1, 2, \dots N$.

To obtain the standard deviation (1-17) and (1-29) are used in (1-28) to yield

$$\sigma = \frac{1}{\sqrt{N-1}} \left[\sum_{j=1}^{N} 10^{(L_j/10)} - N \, 10^{(\bar{L}_p/10)} \right]^{1/2} \qquad (1\text{-}30)$$

It should be noted that σ is *not* expressed as a level in dB. To convert σ to σ_{dB} we note that we can express the deviation about the mean $\bar{p}$ as

$$d = \bar{p} \pm n\sigma = \bar{p}\left(1 \pm \frac{n\sigma}{\bar{p}}\right) \qquad n > 0 \qquad (1\text{-}31)$$

When $n = 1$, $d = \bar{p} + \sigma$ is the value one standard deviation above the mean, and $d = \bar{p} - \sigma$ is the value one standard deviation below the mean. To express (1-31) as a level we take its logarithm. Thus

$$20 \log_{10} d = 20 \log_{10} \bar{p} + 20 \log_{10}\left(1 \pm \frac{n\sigma}{\bar{p}}\right) \qquad (1\text{-}32)$$

(The minus sign in (1-32) can only be used if $n\sigma/\bar{p} < 1$.) The first term on the right-hand side is $\bar{L}_p$. We identify the second term on the right-hand side of (1-32) as σ_{dB} when $n = 1$. Hence one standard deviation above the

*The division by p_{ref} has not been explicitly stated for clarity of presentation.

mean is

$$\sigma_{dB}^{+} = 20 \log_{10}\left(1 + \frac{\sigma}{\bar{p}}\right) \qquad dB \qquad (1\text{-}33)$$

and one standard deviation below the mean is

$$\sigma_{dB}^{-} = 20 \log_{10}\left(1 - \frac{\sigma}{\bar{p}}\right) \qquad dB \qquad (1\text{-}34)$$

provided that $\sigma/\bar{p} < 1$. In (1-33) and (1-34) $\bar{p} = 10^{(L_p/20)}$ and σ is given by (1-30).

In some instances the average sound-power level and its standard deviation are desired. From (1-22) it is seen that in the far-field the sound-power level L_w is proportional to the intensity level L_I plus a constant, and L_I is proportional to the square of the rms pressure [recall (1-19)]. Then the counterpart to (1-27) becomes

$$\left(\overline{p^2}\right) = \frac{1}{N} \sum_{j=1}^{N} p_j^2 \qquad (1\text{-}35)$$

or

$$\overline{L}_w = 10 \log_{10} \frac{1}{N} \sum_{j=1}^{N} 10^{(L_{wj}/10)} \qquad dB \text{ re } 10^{-12} \text{W} \qquad (1\text{-}36)$$

where $\overline{L}_w = 10 \log_{10}(\overline{p^2})$ is the average sound-power level and L_{wj} is the jth sound-power level, $j = 1, 2, \ldots N$.

From (1-28) the standard deviation becomes

$$\hat{\sigma} = \frac{1}{\sqrt{N-1}} \left[\sum_{j=1}^{N} \left(p_j^2\right)^2 - N\left(\overline{p^2}\right)^2 \right]^{1/2} \qquad (1\text{-}37)$$

or

$$\hat{\sigma} = \frac{1}{\sqrt{N-1}} \left[\sum_{j=1}^{N} 10^{(L_{wj}/5)} - N \, 10^{(\overline{L}_w/5)} \right]^{1/2} \qquad (1\text{-}38)$$

Furthermore, one standard deviation above the mean when expressed as a level is

$$\hat{\sigma}_{dB}^{+} = 10 \log_{10}\left(1 + \frac{\hat{\sigma}}{p^2}\right) \qquad dB \qquad (1\text{-}39)$$

and that which is one standard deviation below the mean is

$$\hat{\sigma}_{dB}^- = 10\log_{10}\left(1 - \frac{\hat{\sigma}}{\overline{p^2}}\right) \qquad dB \qquad (1\text{-}40)$$

provided that $\hat{\sigma}/\overline{p^2} < 1$.

Example 1-8. The following eight sound-pressure levels are given : 83.5, 84.2, 88.4, 89.5, 83.7, 85.1, 86.0, and 87.4 dB re 20 μPa. Determine the mean sound-pressure level, its standard deviation, and the sound-pressure level one and two standard deviations from the mean.

Solution: We first determine the following two sums:

$$\sum_{j=1}^{8} 10^{(L_j/20)} = 10^{(83.5/20)} + 10^{(84.2/20)} + 10^{(88.4/20)} + 10^{(89.5/20)}$$

$$+ 10^{(83.7/20)} + 10^{(85.1/20)} + 10^{(86.0/20)} + 10^{(87.4/20)}$$

$$= 16.403 \times 10^4$$

$$\sum_{j=1}^{8} 10^{(L_j/10)} = 10^{(83.5/10)} + 10^{(84.2/10)} + 10^{(88.4/10)} + 10^{(89.5/10)}$$

$$+ 10^{(83.7/10)} + 10^{(85.1/10)} + 10^{(86.0/10)} + 10^{(87.4/10)}$$

$$= 35.756 \times 10^8$$

From (1-29) and (1-30) we obtain

$$\overline{L}_p = 20\log_{10}(16.403 \times 10^4) - 20\log_{10}8 = 86.24 \qquad dB \text{ re } 20\ \mu Pa$$

and

$$\sigma = \frac{1}{\sqrt{7}}[35.756 \times 10^8 - 8 \times 10^{(86.24/10)}]^{1/2} = 5475.0,$$

respectively. Since $\sigma/\overline{p} = 5475.0 \times 10^{-(86.24/20)} = 0.2669$, using (1-33) and (1-34) yields

$$\sigma_{dB}^+ = 20\log_{10}(1 + 0.2669) = 2.055 \text{ dB}$$

and

$$\sigma_{dB}^- = 20\log_{10}(1 - 0.2669) = -2.697 \text{ dB},$$

respectively.

Thus the sound-pressure level one standard deviation above the mean is 86.24 + 2.06 = 88.30 dB and the level one standard deviation below the mean is 86.24 − 2.70 = 83.54 dB. The sound-pressure level two standard deviations from the mean is obtained by recalling (1-32) with $n = 2$. Therefore,

$$\sigma_{dB}^{+} = 20 \log_{10}[1 + (2)(0.2669)] = 3.716 \text{ dB}$$

and

$$\sigma_{dB}^{-} = 20 \log_{10}[1 - (2)(0.2669)] = -6.629 \text{ dB}$$

Thus the sound-pressure level two standard deviations above the mean is 86.24 + 3.72 = 89.96 dB. The sound-pressure level two standard deviations below the mean is 86.24 − 6.63 = 79.61 dB.

Averages and Standard Deviations: Approximate Formulas

Equations 1-29 and 1-30 can be approximated by replacing p_j in (1-27) and (1-28) with L_j, the sound-pressure level. That is, the L_j is not treated as a level but as an ordinal number and (1-27) and (1-28) yield

$$\bar{L}_p^A = \frac{1}{N} \sum_{j=1}^{N} L_j \quad \text{dB} \tag{1-41}$$

and

$$\sigma_{dB}^A = \frac{1}{\sqrt{N-1}} \left[\sum_{j=1}^{N} L_j^2 - N \left(\bar{L}_p^A \right)^2 \right]^{1/2} \text{dB} \tag{1-42}$$

where the superscript A designates the approximate values. How well $\bar{L}_p^A$ and σ_{dB}^A agree with $\bar{L}_p$ and σ_{dB} depends on the magnitude of the difference between the largest L_j and the smallest L_j and the value of N. If this difference is less than 2 dB for sound-pressure levels, the error for the mean will be less than one-tenth of a dB. If the difference is 10 dB the error will be less than 1.4 dB. For sound-power or intensity levels, if the difference is less than 2 dB the error for the mean will be less than 0.17 dB. If the difference is 10 dB, the error will be less than 2.7 dB. These maximum errors will not be exceeded for any value of N. Another error fundamental to (1-42) is that it predicts equal upper and lower standard deviations.

Example 1-9.   Using the data of Example 1-8 determine the approximate mean, standard deviation, and sound-pressure level one and two standard deviations from the mean.

Solution: We first determine the following two sums:

$$\sum_{j=1}^{8} L_i = 687.8$$

$$\sum_{j=1}^{8} L_j^2 = 59169.16$$

Using (1-41) and (1-42) we obtain

$$\overline{L}_p^A = \frac{687.8}{8} = 85.97 \qquad \text{dB re 20 } \mu\text{Pa}$$

and

$$\sigma_{\text{dB}}^A = \frac{1}{\sqrt{7}} \left[59169.16 - 8(85.97)^2 \right]^{1/2} = 2.46 \text{ dB}$$

Thus the sound-pressure levels one standard deviation above and below the mean are 85.97 dB + 2.46 dB = 88.43 dB and 85.97 dB − 2.46 dB = 83.51 dB, respectively. The level two standard deviations above and below the mean are 85.97 dB + (2)(2.46) dB ≒ 90.89 dB and 85.97 dB − (2)(2.46) dB = 81.05 dB, respectively. Hence for the particular values used in Examples 1-8 and 1-9 it is seen that there is a small error introduced by the approximate formula for both the mean and the standard deviation. However, for two standard deviations away from the mean the error is considerably larger.

1.8 OCTAVE BANDS

The human ear is sensitive to sound in the frequency range from roughly 20 Hz to 20 kHz. Since it is almost always impractical to measure the acoustic sound pressure at each frequency in this range, the measurements are made over intervals of frequency. The frequency interval over which measurements are made is called the bandwidth and is specified by an upper and lower frequency limit, f_2 and f_1, respectively, called the cutoff frequencies. In acoustics the frequency bandwidths are usually specified in terms of octaves. An octave is an interval of frequency such that the upper frequency limit is twice the lower limit, that is, $f_2 = 2f_1$. An octave bandwidth may be too large a frequency range in some measurements and a smaller bandwidth may be desirable to use, such as one-third-octave bands.

The general relationship between the upper and the lower cutoff frequencies is given by

$$f_2 = 2^n f_1 \qquad (1\text{-}43)$$

where n is the number of octaves, either a fraction or an integer. For example, $n = 1/3$ specifies a one-third-octave bandwidth, $n = 1/2$ a one-half-octave bandwidth, and $n = 1$ an octave bandwidth. The center frequency f_0 of an n-octave band is the geometric mean of the frequency band given by

$$f_0 = \sqrt{f_2 f_1} \qquad (1\text{-}44)$$

Using (1-43) and (1-44), the upper and lower cutoff frequencies in terms of f_0 and n are found to be

$$f_2 = 2^{n/2} f_0$$
$$\qquad (1\text{-}45)$$
$$f_1 = 2^{-n/2} f_0$$

The bandwidth, B, is

$$B = f_2 - f_1 = f_0(2^{n/2} - 2^{-n/2}) = \beta f_0 \qquad (1\text{-}46)$$

where β is a constant. Hence an n-octave bandwidth is classified as a constant percentage bandwidth, in which the percentage of the center frequency is 100β. For a one-third-octave bandwidth, $n = 1/3$ and $\beta = 0.231$ (23%); for a half-octave bandwidth $n = 1/2$ and $\beta = 0.348$ (35%); and for an octave bandwidth $n = 1$ and $\beta = 0.707$ (70%). The center frequencies and bandwidths for octave and third-octave intervals, which have been standardized by international agreement, are given in Table 1-3.

TABLE 1-3

Center and Approximate Cutoff Frequencies for a Standard Set of Contiguous Octave and 1/3-Octave Band Filters

Band number	Octave Bands			1/3-Octave Bands		
	Approximate lower cutoff frequency (Hz)	Center frequency (Hz)	Approximate upper cutoff frequency (Hz)	Approximate lower cutoff frequency (Hz)	Center frequency (Hz)	Approximate upper cutoff frequency (Hz)
12	11	16	22	14.1	16.0	17.8
13				17.8	20.0	22.4
14				22.4	25.0	28.2
15	22	31.5	44	28.2	31.5	35.5
16				35.5	40.0	44.7
17				44.7	50.0	56.2
18	44	63	88	56.2	63.0	70.8
19				70.8	80.0	89.1
20				89.1	100.0	112.0
21	88	125	177	112.0	125.0	141.0
22				141.0	160.0	178.0
23				178.0	200.0	224.0
24	177	250	355	224.0	250.0	282.0
25				282.0	315.0	355.0
26				355.0	400.0	447.0
27	355	500	710	447.0	500.0	562.0
28				562.0	630.0	708.0
29				708.0	800.0	891.0
30	710	1,000	1,420	891.0	1,000.0	1,122.0
31				1,122.0	1,250.0	1,413.0
32				1,413.0	1,600.0	1,778.0
33	1,420	2,000	2,840	1,778.0	2,000.0	2,239.0
34				2,239.0	2,500.0	2,818.0
35				2,818.0	3,150.0	3,548.0
36	2,840	4,000	5,680	3,548.0	4,000.0	4,467.0
37				4,467.0	5,000.0	5,623.0
38				5,623.0	6,300.0	7,079.0
39	5,680	8,000	11,360	7,079.0	8,000.0	8,913.0
40				8,913.0	10,000.0	11,220.0
41				11,220.0	12,500.0	14,130.0
42	11,360	16,000	22,720	14,130.0	16,000.0	17,780.0
43				17,780.0	20,000.0	22,390.0

2

THE HEARING MECHANISM AND HEARING DAMAGE

2.1 INTRODUCTION

The human ear is the special organ that enables man to hear, that is, to sense and interpret an essentially physical phenomenon—sound. The human ear responds to sound waves in a frequency range from 20 to 16,000 Hz, the upper extremes of which approach 20,000 Hz in the very young. Because the deleterious effects of excessive noise often occur within the ear, some knowledge of the structure and function of the organ is necessary. It is the purpose of this chapter to discuss these aspects of hearing and, in addition, the mechanism of hearing loss. Furthermore, it should provide a rationale for the single number metrics (e.g., A-weighted sound levels) currently in use.

2.2 ANATOMICAL STRUCTURE OF THE HUMAN EAR*

The structure of the human ear is shown in the simplified diagram of Figure 2-1. Because the external ear (not shown in the figure) is small in

*Reprinted, by permission, from *Physical and Applied Acoustics* by Meyer and Neumann, Academic Press, New York, 1972.

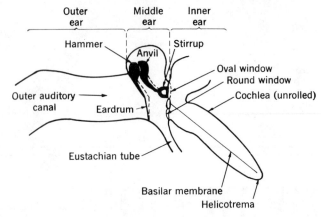

FIGURE 2-1
Schematic of the human ear. (Reprinted, by permission, from *Physical and Applied Acoustics* by Meyer and Neumann, Academic Press, New York, 1972.)

comparison with the wavelengths of the essential components of audible sound, it has only a weak directional effect; however, it does cause the frequency response curve for sound coming from the front of the head to differ from that for sound coming from behind. The auditory canal that begins at the external ear is an acoustically hard tube open at one end and closed at the other end by a compliant membrane called the eardrum. At the midrange frequencies, the eardrum forms an almost reflection-free termination for the auditory canal.

The space behind the eardrum (middle ear) is filled with air and connected by the eustachian tube with the throat. Normally the eustachian tube is closed by the soft palate, so that the sound pressure acts on only one side of the eardrum and one does not hear his own voice too loudly. If the external air pressure changes rapidly, by a change in height, for example, the eustachian tube is opened by involuntarily swallowing, yawning, or the like, to equalize once again the pressure in the middle ear with the external pressure.

In the inner ear, the sound no longer is propagated in air but in a fluid. The transition is provided by the three small bones of the ear—the hammer, the anvil, and the stirrup—which simultaneously transform the particle velocity down and the force up and thus help to match the low characteristic impedance of air to the input impedance of the oval window.

The stirrup is attached to a membrane that forms the oval window of the inner ear. The inner ear consists of a canal filled with lymphatic fluid, coiled up into the shape of a spiral, surrounded by the extremely hard temporal bone. It is almost completely divided along its length into two canals by a mobile wall called the basilar membrane. At the end of the spiral, the two canals are connected by an opening called the helicotrema. Motion of the oval window produces, through the basilar membrane and the helicotrema, a corresponding displacement of the round window.

The basilar membrane is the real organ of reception, for it contains the sensing cells. The motions of the ear drum are transmitted through the small bones of the ear to the membrane of the oval window, which generates a sound wave in the cochlea. The basilar membrane is displaced, and the sensing cells respond to this displacement.

2.3   SOME PROPERTIES OF NORMAL HEARING*

Frequency Range and Sensitivity

If the power of a sound source is gradually reduced, the sound finally becomes so faint that the ear no longer hears it. The threshold of hearing is defined as the sound pressure at which one, listening with both ears in a free field to a signal of waning level, can still just hear the sound, or, if the signal is being increased from a level below the threshold, can just sense it. Measurements made with persons having normal hearing have produced the dashed curve plotted in Figure 2-2 as the average threshold of hearing for pure tones as a function of frequency. The ordinate on the left indicates the rms values of the sound pressure level re 20 μPa and to the right the corresponding sound intensity in dB re 10^{-12} W/m^2. The ear is most sensitive at frequencies from 2000 to 5000 Hz. The smallest perceptible sound pressure in this range amounts to about 20 μPa, which has been adopted as the reference value.

To gain an impression of the great sensitivity of our hearing, recall that atmospheric pressure is approximately 10^5 Pa. The sound pressure 20 μPa corresponds, at 1000 Hz, to a displacement of the air molecules by approximately 0.1 Å. This number is smaller than the mean free path of Brownian molecular motion. It can be estimated that the thermal motion of the air molecules corresponds to a noise sound pressure of about 1 μPa. Thus nature has raised the sensitivity of the ear almost to the physically significant limit. If the ear were much more sensitive, one would hear thermal noise constantly.

*Reprinted, by permission, from *Physical and Applied Acoustics* by Meyer and Neumann, Academic Press, New York, 1972.

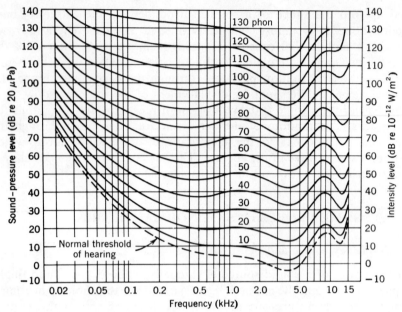

FIGURE 2-2
Equal-loudness contours for binaural free-field listening of pure tones.

Loudness Level

The threshold of hearing is very much frequency dependent. In general, two tones of equal sound pressure but different frequency are not heard as equally loud. Therefore, to characterize a sound by its subjective effect, it is not sufficient to state its intensity level; the characteristics of the ear must be taken into consideration. In addition to the physical quantity, sound pressure or sound-pressure level, a psychological quantity, loudness level, has been introduced.

The definition of loudness level is based on the subjective comparison of two sounds. This is done by sequentially producing, for example, a 100 and a 1000-Hz tone on two loudspeakers standing near each other. When the 1000-Hz loudspeaker is driven at low power, the 100-Hz tone is clearly louder than the 1000-Hz tone; when the 1000-Hz speaker is driven at high power, the reverse is true. In between, the amplifier potentiometer can be set so that the tones will be judged as equally loud, although the pitch and tone quality are quite different. This somewhat difficult hearing comparison is the basis of loudness level determination.

To define loudness level unambiguously, it must first be determined whether the relation "equal loudness" is transferable. If a sound A is adjusted to equal loudness with a sound C and then a sound B is also made equal in loudness to the same sound C, it does not necessarily follow that the sounds A and B would be found equally loud in a direct comparison. Experiment shows, however, that A and B indeed appear to be equally loud within the experimental error.

The loudness level of an arbitrary sound is determined by comparing it to a reference sound of adjustable sound pressure. Standards documents define this reference sound as a plane progressive wave of 1000 Hz. It is adjusted so that it appears equally as loud as the sound being investigated to an observer having normal hearing and listening with both ears to sound incident from in front of him. The sound pressure p of the reference 1000-Hz tone is a measure of the loudness level. It is referred to the sound pressure $p_{ref} = 20$ μPa at the threshold of hearing. Because of the wide dynamic range of the ear, the sound-pressure ratios are expressed logarithmically with respect to the absolute sound-pressure level of the 1000-Hz tone; in other words, the loudness level L of the sound being investigated is defined by the absolute sound-pressure level at which the 1000-Hz tone is judged to be equally loud, $L = 20\log(p/p_{ref})$. To differentiate between this value and the objective sound-pressure level (in decibels), the word "phon" is written after the dimensionless quantity $20\log(p/p_{ref})$.

The numerical value of the loudness level of a 1000-Hz tone is, by definition, equal to the numerical value of its absolute sound-pressure level (see Figure 2-2). The statement "a sound has a loudness level of 60 phon" means simply that it has been found equal in loudness to a 1000-Hz tone of sound level 60 dB, which means a sound pressure of 0.02 Pa or an intensity of 10^{-6} W/m^2.

If the loudness levels of pure tones are measured as a function of frequency, the result is the curves of equal loudness plotted in Figure 2-2 (first obtained by H. Fletcher and W. A. Munson). Considerably higher sound pressures are required to produce the equal-loudness impression at low frequencies than are needed at high frequencies. At the upper end, the lines of constant loudness agree roughly with the lines of constant sound pressure. The threshold of hearing is shown as the lowest curve of equal loudness. The region lying between the threshold of hearing and the threshold of pain in Figure 2-2 is called the hearing area.

2.4   EAR DAMAGE[2]

Introduction

Exposure to noise of sufficient intensity for long enough periods of time can produce detrimental changes in the inner ear and seriously decrease the ability to hear. Some of these changes are temporary and last for minutes, hours, or days after the termination of the noise. After recovery from the temporary effects, there may be residual permanent effects on the ear and hearing that persist throughout the remainder of life. Frequent exposures to noise of sufficient intensity and duration can produce temporary changes that are chronic, though recoverable when the series of exposures finally ceases. Sometimes, however, these chronically maintained changes in hearing lose their temporary quality and become permanent.

The changes in hearing that follow sufficiently strong exposure to noise are complicated. They include distortions of the clarity and quality of auditory experience as well as losses in the ability to detect sound. These changes can range from only slight impairment to nearly total deafness.

Kinds of Ear Damage

Conclusive evidence of the damaging effects of intense noise on the auditory system has been obtained from anatomical methods applied to animals. The inner ears of human beings have also been examined. Some patients with terminal illness have volunteered their inner ears to temporal bone banks. Such specimens are collected at the time of a postmortem examination. The anatomist tries to relate the condition of the human ear to the patient's case history after making allowances for postmortem changes in the inner ear and possible premortem changes associated with the terminal illness or its treatment. In spite of these difficulties, observations of human cochleas are extremely important and in combination with animal experiments provide a fairly clear description of the damaging effects of noise on the inner ear.

Because of the limitations of anatomical methods and the lack of complete knowledge of the relationship between hearing abilities and the anatomy of the auditory system, it is not possible to predict completely the hearing changes from the anatomical changes. However, physiological observations that include measurement of changes in biochemical state and electrical responses of the cochlea and auditory nerve help to reveal the functional changes produced by exposure to noise.

The outer ear, eardrum, and middle ear are almost never damaged by exposure to intense noise, although the eardrum can be ruptured by extremely intense noise and blasts. The primary site of auditory injury from excessive exposure to noise is the receptor organ of the inner ear.

This organ is the organ of Corti, and its normal structure is illustrated in cross-section in Figure 2-3a. Here one can identify the auditory sensory cells (hair cells) and the auditory nerve fibers attached to them, as well as some of the accessory structures of the receptor organ. A brief account of the function of the organ of Corti is as follows. Through a complicated

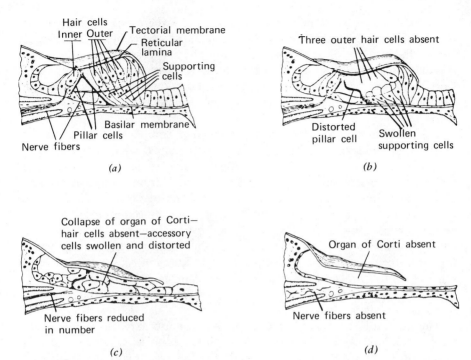

FIGURE 2-3
Human organ of Corti. (*a*) Normal, (*b*) partial injury, (*c*) severe injury, and (*d*) total degeneration. (From Reference 2.)

chain of events, sound at the eardrum results in an up-and-down movement of the basilar membrane. The hair cells are rigidly fixed in the reticular lamina of the organ of Corti which, in turn, is fixed to the basilar membrane. As the basilar membrane is driven up and down by sound, a shearing movement is generated between the tectorial membrane and the top of the organ of Corti. This movement bends the hairs at the top of the hair cells. This bending, in turn, causes the hair cells to stimulate the auditory nerve fibers. As a result, nerve impulses arise in the nerve fibers and travel to the brain stem. From the brain stem the nerve impulses are relayed to various parts of the brain and in some unknown way give rise to auditory sensations. The point to be made is that the integrity of the sensory cells and the organ of Corti is important for normal hearing.

Excessive exposure to noise can result in the destruction of hair cells and the collapse or total destruction of sections of the organ of Corti. In addition, auditory neurons may degenerate. Figure 2-3 illustrates these injuries. The injury illustrated in Figure 2-3*b* includes the absence of three outer hair cells, distortion of a pillar cell, and swelling of the supporting cells. In Figure 2-3c there is a complete collapse of the organ of Corti as well as the absence of hair cells, distortion of the accessory structures, and a reduction in the number of nerve fibers. This section of the organ of Corti is almost certainly without auditory function. The injury shown in Figure 2-3d is obvious; there is complete degeneration of the organ of Corti.

The loss of hearing abilities depends, in a complicated way, on the extent of the injury along the organ of Corti. Total destruction of the organ of Corti for 1 or 2 mm of the total 34 mm may or may not lead to measurable changes in hearing. Recent evidence from human cases and animal experiments suggests that the loss of sensory cells must be quite extensive in the upper part of the cochlea (which is important for the perception of low-frequency sounds) before this damage is reflected as a change in threshold. In the lower part of the cochlea (which is important for the perception of high-frequency sounds) losses of sensory cells over a few millimeters are sometimes reflected in changes in hearing. The loss of 15 to 20% of the cells that respond to high frequencies can reduce hearing sensitivity by as much as 40 dB.

The mechanism by which over-exposure to noise damages the auditory receptor is not well understood. Very intense noise can mechanically damage the organ of Corti. Thus loud impulses, such as those associated with explosions and firing of weapons, can result in such severe vibrations of the organ of Corti that some of it is simply torn apart. Other very severe exposures to noise may cause structural damage that leads to rapid "break-down" of the processes necessary to maintain the life of the cells of the organ of Corti. Such an injury is an acoustic trauma.

Overexposure to noise of lower levels for prolonged periods of time also results in the degeneration of the hair cells and accessory structures of the organ of Corti. Such injuries are called noise-induced cochlear injuries. Many theories have been proposed to explain noise-induced cochlear injuries. One notion is that constant overexposure forces the cells to work at too high a metabolic rate for too long a period of time. As a result, the metabolic processes essential for cellular life become exhausted or poisoned, and this leads to the death of the cells. In a sense, the receptor cells can die from overwork.

No matter what theory is eventually found to be correct, certain facts are established beyond doubt. Excessive exposure to noise leads to the destruction of the primary auditory receptor cells, the hair cells. Other injuries to the organ of Corti can range from mild distortion of its structure to collapse or complete degeneration. The auditory neurons may also degenerate. All of these cells are highly specialized and once destroyed they do not regenerate and cannot be stimulated to regenerate; they are lost forever.

2.5 HEARING LOSS—THRESHOLD SHIFTS*

The primary measure of hearing loss is the hearing threshold level. As previously stated the hearing threshold level is the level of a tone that can just be detected. The diminution, following exposure to noise, of the ability to detect weak auditory signals is termed temporary threshold shift (TTS), if the decrease in sensitivity eventually disappears, and noise-induced permanent threshold shift (NIPTS) if it does not. For years it has been assumed that these two phenomena were very closely related: (1) that noises producing equal average amounts of TTS (sometimes called auditory fatigue) would also produce equal amounts of NIPTS; (2) that if one noise produced twice as much TTS as another, it was also twice as dangerous in regard to NIPTS; and (3) that if one individual showed half as much TTS as another, he would suffer only half as much permanent loss. However, current research leads to the tentative conclusion that the characteristics most important in determining whether an ear will get a relatively large amount of TTS are not also those that determine the degree of final loss from a particular exposure, even when the spectrum of the noise is constant.

Nevertheless, known relations specifying the effect of various characteristics of noise on TTS constitute some evidence that supports inferences about the relative noxiousness of different noise exposures in regard to NIPTS. Such relations have in fact been used in deriving the recent damage-risk criteria.

The most firmly established relations between noise and TTS are:

1. The growth of TTS in dB is nearly linear with the logarithm of time. Moderate TTS also recovers exponentially in time, recovering completely within 16 hr after exposure. However, when the TTS has reached 40 dB or more, recovery may become linear in time, with the TTS requiring days or even weeks to disappear. This 40 or 50 dB of TTS may represent some sort of "critical TTS" that should not be exceeded if danger of permanent damage is to be avoided.

2. Noises whose maximum energy is in low frequencies will produce less TTS than those whose energy is at higher frequencies. That is, a rumble is less dangerous than a screech.

3. The maximum effect from a noise that has energy concentrated in a

*Reprinted, by permission, from Reference 3.

narrow frequency range will be found half an octave to an octave above that range rather than in it.

4.   TTS increases linearly with the average noise level, beginning at about 80 dB sound-pressure level, at least up to 130 dB or so. In other words, the difference between TTSs produced by 100- and 110-dB noises will be about the same as the difference between those produced by 110 and 120 dB.

5.   An intermittent noise is much less able to produce TTS than a steady one. In fact, the TTS is proportional to the fraction of the time that the noise is present. A noise that is on only half the time (in bursts of a few minutes or less) can be tolerated for much more than twice the number of working hours than could be spent in the noise when it is continuous, before the same TTS would be produced.

6.   Neither growth nor recovery of TTS is influenced by drugs, medications, time of day, hypnosis, good thoughts, or extrasensory perception. Thus the locus of the physiological deficit associated with TTS seems to be extremely peripheral—at the hair cells of the cochlea, to be specific.

There are many misconceptions concerning NIPTS. Below, in question and answer form, some of the more important ones are dispelled.

1.   *Are certain frequencies more sensitive than others to damage from noise*? After long exposure to industrial noise or, for that matter, to gunfire, the frequencies showing first and most severe NIPTSs are those in the vicinity of 4000 Hz, with neighboring frequencies affected later. The reason for this seems to be a combination of two factors: (1) the middle ear transmits the frequencies between 1000 and 4000 Hz most efficiently, so that more energy reaches the cochlea in this range, and (2) a given area of the basilar membrane is affected by a wide range of frequencies below its characteristic frequencies, but not by those above; therefore, many of the most intense noise elements affect the 4000 Hz receptors.

2.   *How long must the ear be out of noise before it will have recovered all it is going to*? Two weeks is mandatory but little further recovery occurs after a month, although occasionally, following trauma from a single incident (such as a firecracker exploding near the ear), slight additional recovery may occur in the second month.

3.   *Is NIPTS a progressive process in the sense that, once started, it continues even though the individual is removed from the noise*? Although many people still suspect that this may be so, the evidence is always equivocal. When the hearing of a group of people who have been removed from noise is followed over a period of years, there are always a few who show slight additional losses. However, whether or not the amount of increase is greater than what would be expected in any group of in-

dividuals (i.e., whether or not the additional loss is merely sociocusis that occurs because the total acoustic environment of the ears during the intervening years cannot be controlled) is generally disputed. There is as yet no convincing proof that any progressive degenerative process is set in motion.

4. *But is the noise-damaged ear more susceptible to further injury than a normal ear?* The difficulty in answering this question arises from the difficulty of equating injury to normal and to already-damaged ears. Is a 10-dB increase in PTS in an ear that already had a 40-dB loss smaller, equal, or greater than a 20-dB change in an ear that was initially "normal"? Numerically, it is smaller. But it represents a greater loss of loudness in a normal ear. Again there seems to be no evidence that an ear with some NIPTS is more susceptible than a normal ear, particularly if all temporary effects have completely disappeared.

5. *If permanent injury does not occur, does habitual exposure to a moderate noise render the ear more resistant to an occasional high-intensity exposure? That is, does the ear get "tougher"?* There is no evidence that the basilar membrane will become more leathery, or that the middle-ear muscles, which presumably help to protect the inner ear, become stronger as time goes on. In fact, it has recently been found that just the opposite happens: 15- to 18-year-old boys allegedly showed more auditory fatigue after working in noise for several months than they did at the beginning of employment.

6. *Are there any ameliorative agents that will decrease the PTS produced by a given noise?* One can do little to inhibit the growth of PTS or to cure it. For a while, there was hope that massive doses of vitamin A might reduce NIPTS but subsequent studies failed to confirm an action of vitamin A on either TTS or PTS. Biochemists are studying the effect on TTS and PTS of a broad spectrum of agents but no clear effect has been demonstrated. Rather than admit that there is no effective therapy for an existing NIPTS, some physicians still recommend stellate blocking, novocaine, hydergin, vasodilators, and vitamins, but placebos would probably do as much good.

2.6 DAMAGE RISK CRITERIA

Introduction

Studies* carried out in different parts of the world have been consistent in predicting both temporary and permanent hearing loss due to excessive

*See, for example, Reference 4.

exposure to noise. Regardless of the source of information, the following conclusions are generally accepted: (1) The most important function of human hearing is to hear and understand speech. (2) Significant difficulty in hearing speech (hearing handicap) is not experienced until the average hearing threshold levels at 500, 1000, and 2000 Hz exceed 25 dB [relative to the accepted audiometric reference level—see American National Standards Specification for Audiometers (S3.6-1969)]. (3) Noise levels below an A-weighted* sound level of 80 dB will not produce, for 90% of the populace, a change in hearing sufficient to cause difficulty in speech reception. (4) The amount of hearing loss produced by levels higher than this is a function of noise level, time distribution of noise exposure during the exposure period, total duration of the exposure over a lifetime, and individual susceptibility.

As long as a person hears well in the range 350-3000 Hz, he will have no trouble understanding speech. Generally, perception of the higher frequencies is the first to be affected by excessive noise, so that a person may be unaware that hearing damage is taking place. High frequency hearing loss may make children's voices very difficult or impossible to understand, and ringing of the telephone may be muffled. As loss progresses to lower frequencies, rapid speech, particularly in a noisy background, will be very hard to understand. If damage continues, understanding of normal conversation becomes difficult or impossible.

As indicated previously, a measure of hearing loss is the average hearing threshold level, sometimes abbreviated HL, averaged [as per (1-41)] for 500, 1000, and 2000 Hz in the better ear. If the HL is less than 25 dB, that is, if the minimum audible loudness contour of Figure 2-2 is essentially shifted upwards less than 25 phons, a person will have no significant difficulty understanding faint speech. If the HL is between 25 and 40 dB, difficulty will be encountered with faint speech. A person with a HL between 40 and 55 dB will encounter frequent difficulty with normal speech whereas a HL between 55 and 70 dB will cause frequent difficulty with loud speech. For a HL between 70 and 90 dB only amplified speech can be understood. For HLs greater than 90 dB even amplified speech will not be understood.

Noise-Induced Hearing Loss and Risk of Impairment

The manner in which permanent threshold changes occur as a result of continuous noise exposure is shown in Figure 2-4. The data[5] used to develop these curves were obtained from HL data averaged for both ears

*See p. 49 for the definition of the A-weighted sound level.

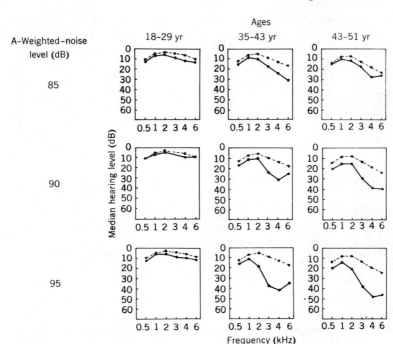

Ages

A-Weighted-noise
level (dB)

18-29 yr   35-43 yr   43-51 yr

85

90

95

Frequency (kHz)

FIGURE 2-4
Occupational hearing level distribution grouped by age and A-weighted noise level.
Non-noise group (-----). Noise group (⎯⎯⎯). (From Reference 5.)

of 1378 employees from 13 different companies comprising 9 different industries. This figure clearly demonstrates the effect of noise upon hearing, particularly at frequencies of from 2000 to 6000 Hz. Similar hearing losses occur when workers are exposed to impulsive type noises. Figure 2-5 shows the hearing damage to workers in a drop-forge plant.[6] The background noise levels for these workers was 110 dB and the peak pressure levels were between 127 and 134 dB. Each worker was exposed to from 3000 to 10,000 impulses a day. The duration of the impulse ranged from 100 to 200 msec.

Other data,[7] shown in Figure 2-6, indicate the percentage of the population that will sustain a noise-induced permanent hearing loss (as defined on p. 37) after a number of years of exposure for 8 hr per day. The number of years of exposure may be computed by subtracting 18 yr from the age. From this figure it is clearly seen how the risk of impaired hearing increases as the levels of continuous noise exposure increase beyond an A-weighted noise level of 85 dB.

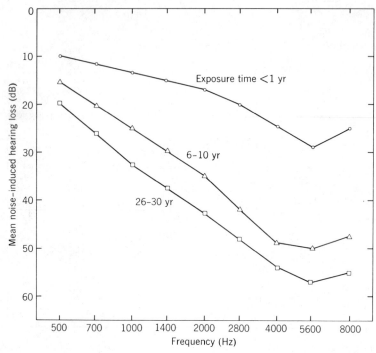

FIGURE 2-5
Mean impulsive noise-induced hearing loss as a function of years of exposure. (From Reference 6.)

Acceptable Noise Exposures[8]

All day exposures to steady noise have been investigated by retrospective and prospective studies that indicate the noise levels at which hearing damage begins after many years of redundant exposure. Part-day exposures have been studied less because of the greater complexities involved. As a result, theories are relied upon to set limits for part-day exposures. The currently accepted theory is the equal temporary effect theory.

The equal temporary effect theory postulates that the hazard of noise exposure increases with the average temporary loss of hearing it produces in young normal ears. This theory is immediately appealing since it is related to the function of the ear. It arises out of the observation that those noise exposures that ultimately produce permanent hearing loss also produce temporary hearing loss in young normal ears. Conversely, those noise exposures that do not produce permanent hearing loss likewise do not produce temporary hearing loss in young normal ears. Although a causal

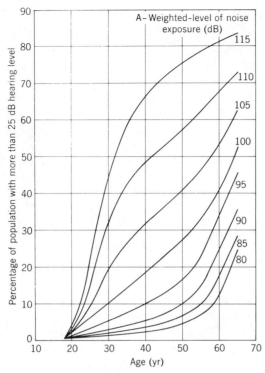

FIGURE 2-6
Percentage of population with more than 25 dB hearing level as a function of exposure to various A-weighted levels. (From Reference 7.)

relation between temporary hearing loss and permanent hearing loss has not been established, the results of temporary threshold shift studies have been used to define safe limits for all-day exposures to steady noise which agree with those established by permanent threshold shift studies. Studies of TTS also lead to reasonable limits for very short or indefinitely long exposures. TTS studies indicate that intermittent noise is much less harmful than steady noise, which is supported by all current evidence.

As previously mentioned, the relationship between TTS and NIPTS is not very good. However, researchers have noted that a TTS 2 min after the end of an 8-hr exposure to three types of noise (denoted TTS_2), at a test frequency of 4000 Hz, was the same as the asymptotic NIPTS following 10–27.5 yr of daily exposure to each of these noises. The observed relations suggested that noise exposures could be ranked in severity according to the amount of TTS they produced. There was even the possibility that the

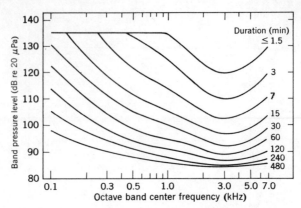

FIGURE 2-7
Octave band levels at which an exposure for the duration indicated will produce
average TTSs equal to the damage risk criteria. (Reprinted, by permission, from
Reference 8.)

amount of TTS_2 would be reasonably predictive of the amount of NIPTS.
From these observations a set of contours, called equally harmful contours, was developed. This is shown in Figure 2-7 in terms of octave-band
level and center frequency of the octave bands, with duration of exposure
as a parameter. For example, the TTS_2 produced by an octave band of
noise centered at 2 kHz when the level of exposure is 85 dB and the
duration is 480 min should be equal to that at 88 dB for 120 min, to that at
96 dB for 30 min, or to that at 106 dB for 7 min. The exposure described
by each contour will, on the average, produce the criterion TTS_2. Decreases of either noise level or duration of exposure will result in less TTS
and increases will result in more. This figure emphasizes the spectral
dimensions of the relations and shows that for long durations the contour
for octave-band level does not vary much with frequency, but for short
durations low-frequency noises up to 130 dB do not produce criterion TTS.

The results shown in Figure 2-7 require octave band sound-pressure
levels. Botsford[9] rearranged the relations shown by these contours by
examining the spectral composition of commonly encountered manufactur-
ing noises and replacing the specifications in terms of octave-band levels
with specifications in terms of A-weighted* sound-pressure levels.

As described in Section 2.7, the results of this section are the basis for
the government-imposed limits set by the Walsh-Healey Act.

*See p. 49.

2.7 THE WALSH-HEALEY ACT

The Bureau of Labor Standards, U. S. Department of Labor, added a safety regulation on industrial noise exposure to the Walsh-Healey Act in May 1969. The levels and criteria stated below are based, in part, on the results of Section 2.6. The regulation, in part, stipulates that:

1. Protection against the effects of noise exposure will be provided when sound levels exceed those shown in Table 2-1 when measured on the A scale of a standard sound level meter at slow response.

TABLE 2-1
Permissible Noise Exposure

Duration per day (hr)	A-weighted sound level (dB)
8.0	90
6.0	92
4.0	95
3.0	97
2.0	100
1.5	102
1.0	105
0.5	110
0.25 or less	115 (maximum)

2. When employees are subjected to sound levels exceeding those listed in Table 2-1, feasible administrative or engineering controls will be utilized. If such controls fail to reduce sound levels to within the levels of the table, personal protective equipment will be provided to reduce them to these values.

Regarding the usage of Table 2-1, when the daily noise exposure is composed of two or more periods of noise exposure of different levels, their combined effect should be considered, rather than the individual effect of each. Let C_n indicate the total time of exposure at a specified noise level and T_n indicate the total time of exposure permitted at that level. If the sum $\Sigma\, C_n/T_n$ exceeds unity, the mixed exposure has exceeded the maximum permissible duration.

2.8   DAMAGE RISK FOR IMPULSIVE NOISE

Guidelines have been prepared to estimate the hearing damage risk associated with impulse noise, such as gunfire. Impulse noises, for this purpose, can be defined as brief noises less than 1 sec.

The available data on hearing damage due to impulse noises are mainly derived from studies of military personnel. In general these data do not provide reliable indications of the actual noise exposures that caused measured hearing loss. The guidelines discussed below are based primarily on results of temporary threshold shift studies. Since there are essentially no data directly relating TTS from a single noise exposure to the noise-induced permanent threshold shift from habitual exposure, these recommendations should be used with some caution. However, they do represent the best information available to date.

Impulse noises are divided into the two general types illustrated in Figure 2-8, although intermediate forms do occur. Figure 2-8a shows the pressure waveform that is often observed when a gun is fired outdoors with no reflecting surfaces nearby. Figure 2-8b exemplifies a much more complicated situation: an initial series of damped oscillations which may be followed by a reflected wave at only a slightly lower level. There are three parameters of a single impulse noise that are of importance to the criterion being discussed:

1. The *peak pressure level* (P) is the highest instantaneous pressure level (in dB re 20 μPa) reached at any time by the impulse, measured at the position of the ear with the individual not present.

2. The *pressure-wave duration*, or *A-duration*, is the time for the initial or principal pressure wave to rise to its positive peak and return to ambient pressure. In the ideal pressure wave shown in Figure 2-8a, the A-duration is given by the distance ($W - V$) on the time axis.

3. The *pressure-envelope duration*, or *B-duration*, is the total time that the envelope of the pressure fluctuations, both positive and negative, is within 20 dB of the peak pressure level. Included in this time is the duration of that part of any reflection pattern within 20 dB of the peak level. Thus in Figure 2-8b, the B-duration is given by ($X - V$)+($Z - Y$).

Figure 2-9 presents the fundamental criterion developed by the CHABA working group.[10] This criterion is intended to limit the temporary threshold shift, produced in all but the most susceptible 5% of exposed individuals, measured 2 min after cessation of exposure to the noise, to less than 10 dB at 1000 Hz or above. The criterion is based on the assumption that the permanent hearing loss (NIPTS) eventually produced by many years of exposure to the same noise is approximately equal to the temporary threshold shift shown by a normal ear after a single day's exposure to the

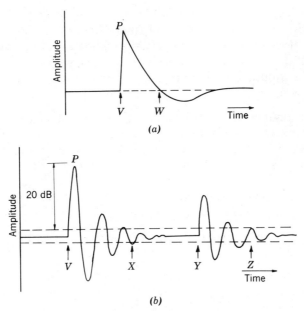

FIGURE 2-8
Two principal types of impulse noise: (*a*) A-duration and (*b*) B-duration.

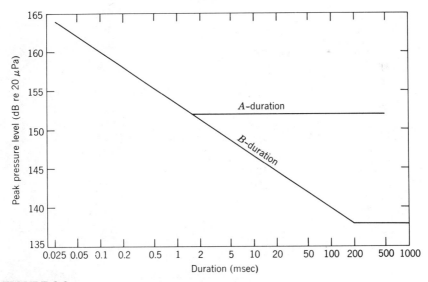

FIGURE 2-9
Damage risk criterion for impulse noise based on 100 exposures per day. See text foɪ
corrections for other than 100 exposures per day.

same noise. The criterion shown in Figure 2-9 represents the limits for 100 impulses distributed over a period of from 4 min to several hours on any single day. It is assumed that the pulses reach the ear at normal incidence. In case of doubt as to which waveform analysis to apply, the more conservative *B*-duration should be used. The main features of the criterion are as follows:

1. The maximum peak pressure level permitted, without ear protection, is 164 dB for the shortest pulse of any practical interest (25 μsec).
2. As duration increases, the permitted peak pressure level decreases steadily at a rate of 2 dB for each doubling of the duration, dropping to a terminal level of 138 dB for *B*-durations of 200–1000 msec.
3. A similar decrease occurs for *A*-durations, except that terminal level of 152 dB is reached at about 1.5 msec.

In case the conditions stipulated for this basic criterion are not met, correction factors can be applied as follows:

1. If the pulses arrive at the ear at grazing incidence instead of at normal incidence, the curves can be shifted upward by 5 dB, that is, 5 dB can be added to the ordinate values in Figure 2-9.
2. If the number of pulses in an "exposure period" (i.e., on any given day) is some value other than 100, the following corrections should be added to the permissible peak pressure level of the impulse:

Number of impulses per day	Correction (dB)
1	+15
10	+11
20	+10
50	+ 5
100	0
200	− 5
300	− 10
500	− 15
1000	− 20
2000	− 25
5000	− 30

REFERENCES

1. E. Meyer and E. G. Newmann, *Physical and Applied Acoustics*, Academic Press, N. Y. (1972), pp. 234–236, 241–248.

2. J. D. Miller, "Effects of Noise on People," Report No. NTID 300.7, U. S. Environmental Protection Agency, Washington, D. C. (December 1971).

3. W. D. Ward, "Effects of Noise on Hearing Thresholds," in *Noise as a Public Health Hazard*, W. D. Ward and J. E. Fricke, (Eds.), American Speech and Hearing Association, Washington, D. C. (February 1969), pp. 40–48.

4. W. D. Ward, (Ed.), "Proceedings of the International Congress on Noise as a Public Health Problem," Dubrovnik, Yugoslavia, May 13–18, 1973, Report No. 550/9-73-008, U. S. Environmental Protection Agency, Washington, D. C.

5. B. L. Lempert and T. L. Henderson, "Occupational Noise and Hearing 1968 to 1972," Report No. NIOSH-TR-201-74, U. S. Department of Health, Education, and Welfare, National Institute for Occupational Safety and Health, Cincinnati, Ohio (1973). (NTIS No. PB-232-284.)

6. T. Ceypek and J. J. Kuźniarz, "Hearing Loss Due to Impulse Noise; A Field Study," Ref. 4, pp. 219–228.

7. W. L. Baughn, "Relation Between Daily Noise Exposure and Hearing Loss Based on the Evaluation of 6,835 Industrial Noise Exposure Cases," Report No. AMRL-TR-73-53, Aerospace Medical Research Laboratory, Wright-Patterson Air Force Base, Ohio (June 1973).

8. D. H. Eldridge and J. D. Miller, "Acceptable Noise Exposures-Damage Risk Criteria," Ref. 3, pp. 110–120.

9. J. H. Botsford, "Damage Risk," in *Transportation Noise*, J. D. Chalupnik, (Ed.), University of Washington Press, Seattle, Washington (1970), pp. 103–113.

10. W. D. Ward et al., "Proposed Damage-risk Criterion for Impulse Noise (Gunfire)," Report of Working Group 57, NAS-NRC Committee on Hearing, Bioacoustics, and Biomechanics (CHABA) (July 1968).

3

PSYCHOLOGICAL
AND
SOCIOLOGICAL
INTERPRETATION
OF SOUND

3.1 INTRODUCTION

Noise, in environmental acoustics, refers to sound that is unwanted by the listener, presumably because it is unpleasant or bothersome, it interferes with the perception of the wanted sound, or it is physiologically harmful. Noise, an unwanted sound, does not necessarily have any particular physical characteristic to distinguish it from wanted sound. Therefore, the word "noise" is often used in place of "sound" merely to draw attention to its "unwantedness."

There are certain unwanted effects of sounds that appear to be related rather precisely to physical characteristics of sound in ways that are more or less universal and invariant for all people. These effects are (a) masking of wanted sounds, particularly speech, (b) auditory fatigue and damage to

48

hearing, (c) excessive loudness, (d) some general quality of bothersomeness or noisiness, and (e) startle. Some of these universal aspects of noise are discussed in this chapter and the means whereby they can be quantified to predict the reaction of groups of people.

We also consider numerous different rating scales and procedures intended for the evaluation of noise according to the predicted subjective response to that noise. The distinction between "scales" and "procedures" is as follows. A scale describes only the noise itself, either in a simple way, such as the reading of the maximum A-weighted sound level during a transient noisy·event, or in some more complicated manner that may account for the time variation of noise analyzed into frequency bands, perhaps in terms of the statistical distribution of instantaneous levels regarded as a time series. This basic acoustical information may form the basis of a computed scale that purports to predict some particular subjective response, individual or group-average, to that noise, for example, the loudness, the judged noisiness, or the interference with oral communication. In any case, a scale attempts only to describe some aspect of the noise stimulus itself. The rating scales discussed in this chapter are: A-weighted-sound-pressure level, Speech Interference Level (SIL), Loudness Level (LL), Perceived Noise Level (PNL), Noise Criterion (NC), Traffic Noise Index (TNI), and Noise Pollution Level (NPL).

A rating procedure, on the other hand, attempts to account for the context in which the noise stimulus is experienced and usually entails the introduction of corrections, both for peculiar characteristics of the noise (pure tone content, duration, or impulsive or intermittent nature) and for the situation into which it intrudes (type of neighborhood, time of day, week, or year, etc.). The rating procedures evaluated in this chapter are: the day-night sound level (L_{dn}) and the Noise Exposure Forecast (NEF).

3.2   THE A-WEIGHTED-SOUND LEVEL

The A-weighted-sound level is the level, in dB re 20 μPa, of a sound impinging upon a microphone that has been electronically altered with a weighting network (filter) whose frequency response is called the A-weighting curve. The overall characteristics of this A-weighting curve are approximately equal to the Fletcher-Munson type equal-loudness contour for a loudness level of 40 phons. Figure 3-1 illustrates a plot of the A-weighting curve as a function of frequency. Two other weighting networks are sometimes used: the B-weighting and the C-weighting. The B-weighting corresponds approximately to a loudness contour of 70 phons and the C-weighting to that of 100 phons. These contours are also shown

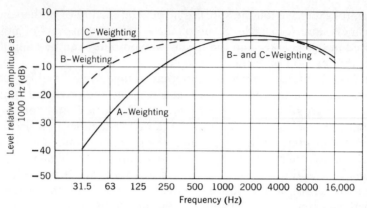

FIGURE 3-1
Internationally standardized A-, B- and C-weighting curves.

in Figure 3-1. Since 1/3- and 1/1-octave data are often converted to an A-weighted-sound level, Table 3-1 lists the weights at the center frequencies of the 1/3-octave bands. It should be apparent that the A-weighted level is a broad-band measurement, and as is the case in all broad-band measurements, sounds with different power spectra can have the same value. This is shown in Figure 3-2. The A-weighted level persists in correlating almost as well with subjective evaluations of many varieties of noise as the most sophisticated scales that purport to account for the

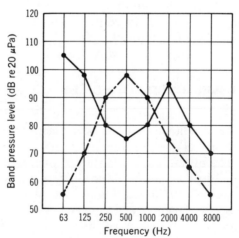

FIGURE 3-2
Two different sounds with the same A-weighted-sound level.

frequency variable. It should be cautioned, however, that the A-weighted level is used for broad-band noise sources. It should not be used for narrow-band noises or noises with discernable pure tones.

TABLE 3-1
A-Weighting Response as a Function of 1/3-Octave Band Center Frequencies[a]

1/3-Octave band center frequency (Hz)	A-Weighting response (dB re amplitude at 1000 Hz)	1/3-Octave band center frequency (Hz)	A-Weighting response (dB re amplitude at 1000 Hz)
10	− 70.4	500	− 3.2
12.5	− 63.4	630	− 1.9
16	− 56.7	800	− 0.8
20	− 50.5	1,000	0
25	− 44.7	1,250	+ 0.6
31.5	− 39.4	1,600	+ 1.0
40	− 34.6	2,000	+ 1.2
50	− 30.2	2,500	+ 1.3
63	− 26.2	3,150	+ 1.2
80	− 22.5	4,000	+ 1.0
100	− 19.1	5,000	+ 0.5
125	− 16.1	6,300	− 0.1
160	− 13.4	8,000	− 1.1
200	− 10.9	10,000	− 2.5
250	− 8.6	12,500	− 4.3
315	− 6.6	16,000	− 6.6
400	− 4.8	20,000	− 9.3

[a]These values assume that the microphone and the electronics following have a flat, diffuse-field response. See Section 4.5.

As mentioned above, 1/3- and 1/1-octave band levels are often converted into an A-weighted level. This conversion is performed with the following relation obtained from (1-24):

$$L_A = \text{A-weighted level} = 10\log_{10} \sum_{j=1}^{N} 10^{\{(L_j + A_j)/10\}} \qquad \text{dB re } 20 \ \mu\text{Pa} \quad (3\text{-}1)$$

where L_j are the 1/3- or 1/1-octave band levels and A_j are the A-weights obtained from Table 3-1 at the center frequencies of either the 1/3- or 1/1-octave bands.

Example 3-1.   Verify that the two curves shown in Figure 3-2 have the same A-weighted levels.

Solution: Dashed curve:

$$A\text{-weighted level} = 10\log_{10}[10^{(55-26.2)/10} + 10^{(70-16.1)/10} + 10^{(90-8.6)/10}$$

$$+ 10^{(98.5-3.2)/10} + 10^{90/10} + 10^{(75+1.2)/10}$$

$$+ 10^{(65+1.0)/10} + 10^{(55-1.1)/10}]$$

$$= 96.6 \text{ dB re } 20 \ \mu\text{Pa}$$

Solid curve:

$$A\text{-weighted level} = 10\log_{10}[10^{(105-26.2)/10} + 10^{(98-16.1)/10} + 10^{(80-8.6)/10}$$

$$+ 10^{(75-3.2)/10} + 10^{80/10} + 10^{(95+1.2)/10}$$

$$+ 10^{(80+1.0)/10} + 10^{(70-1.1)/10}]$$

$$= 96.7 \text{ dB re } 20 \ \mu\text{Pa}$$

Thus the A-weighted-sound levels of the two noises with different octave band spectra are within less that 0.1 dB of each other.

3.3   SPEECH INTERFERENCE LEVEL (SIL)[1-3]

The Speech Interference Level was devised by Beranek as a simplified substitute for the Articulation Index* and originated mainly to evaluate the cabin noise in aircraft, where it has been most widely used. The SIL is the arithmetic average of the sound pressure levels in the three octave bands lying between 600 and 4800 Hz and is only a measure of the noise background. One can refer to a table indicating, for each value of SIL, the voice level required to communicate reliably over various talker-listener distances.

*The Articulation Index's basic measure is a weighted signal-to-noise ratio, based on normal speech levels and measured (or estimated) background noise levels. It is, in effect, a weighted fraction representing, for a given speech channel and noise condition, the proportion of the normal speech signal available to a listener for conveying speech intelligibility. The articulation index yields an accurate prediction, from purely physical measures of a communications channel, of the intelligibility of speech transmitted over that system.

With adoption, in 1960, of a new set of preferred frequencies for analysis, measured octave-band data gradually became incompatible with Beranek's original definition. After experimentation with several proposed replacements, the average of the levels in the octave bands centered on 500, 1000, and 2000 Hz has been recommended as the Preferred Speech Interference Level (PSIL); it appears to be gaining acceptance for the same purpose as the original SIL.

Relations among PSIL, voice effort, and background noise level are shown in Table 3-2. The values of the PSIL are for steady continuous noises at which reliable speech communication is barely possible between persons at the distances and voice efforts shown. The interference levels are for average male voices (reduce the levels 5 dB for female voices) with speaker and listener facing each other, using unexpected word material. It is assumed that there are no nearby reflecting surfaces that aid the speech sounds.

TABLE 3-2
Relations Among PSIL, Voice Effort, and Background Noise[a]

| Distance between talker and listener (m) | PSIL[b] (dB) | | | |
| | Talker's voice effort | | | |
	Normal	Raised	Very loud	Shouting
0.15	74	80	86	92
0.3	68	74	80	86
0.6	62	68	74	80
1.2	56	62	68	74
1.8	52	58	64	70
3.7	46	52	58	64

[a]From *Noise and Vibration Control* edited by L. L. Beranek. Copyright 1971 by McGraw-Hill Book Company. Used with permission of McGraw-Hill Book Company.
[b]SIL (calculated from old octave bands) $\approx$ PSIL -3 dB.

It is often more convenient to talk in terms of speech interference and A-weighted background noise levels. This is depicted in Figure 3-3. The vertical axis is the A-weighted sound· level of background noise. The horizontal axis is the distance, measured in meters, between talker and listener. The regions below the contours are those combinations of distance, background noise levels, and vocal outputs wherein speech communication is practical between young adults who speak similar dialects of American-English. The line labeled "expected voice level" reflects the fact that the usual talker unconsiously raises his voice level when he is surrounded by noise. Consider the example of a talker who wishes to speak to a listener near a running faucet. The A-weighted-sound level of the background noise may be about 74 dB for the listener. If the talker is 6 m away, it is clear from Figure 3-3, as well as from everyday experience, that communication would be difficult even if the talker were to shout. But if the talker were to move within 0.3 m of the listener, communication would be practical even when a normal voice is used. It can be seen that at 4.5–6 m, distances not uncommon to many living rooms or classrooms, A-weighted-sound levels of the background noise must be below 50 dB if speech communication is to be nearly normal.

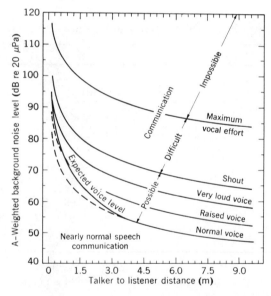

FIGURE 3-3
Quality of speech communication as a function of the A-weighted-sound level of the background noise and the distance between talker and listener. (From Reference 3.)

A third way[3] of presenting speech intelligibility data is shown in Figure 3-4 as a function of three different vocal efforts, steady A-weighted background noise levels, distance from speaker to listener, and percentage sentence intelligibility. This latter quantity indicates the percentage of words spoken that could be correctly understood. Ninety-five percent sentence intelligibility usually permits reliable communication because in most normal conversations the vocabulary is limited and the few unheard words can be inferred.

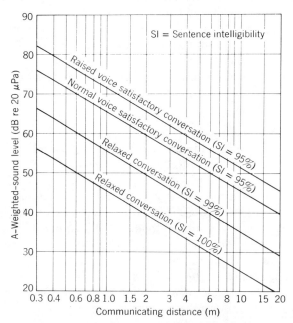

FIGURE 3-4
Maximum distances over which conversation is considered satisfactorily intelligible in the presence of a continuous noise. (From Reference 3.)

People vary their voice levels and distances not only in accordance with the level of background noise and physical convenience, but also in accordance with cultural standards. Distances less than about 1.4 m are reserved for confidential or personal exchanges usually with a lowered voice. Distances greater than about 1.5 m are usually associated with a slightly raised voice and are reserved for messages that others are welcome to hear. Thus levels of background noise that require the talker and listener to move within less than 1.2 m will be upsetting to persons who do not normally have an intimate association. Even for close friends there may be some embarrassment if the message would not normally require such nearness. When the content of the message is personal, there will be reluctance to raise the voice even if the background noise demands it for intelligibility.

In one-to-one personal conversations the distance from talker to listener is usually about 1.5 m and nearly normal speech communication can proceed in A-weighted-noise levels as high as 66 dB. Many conversations involve groups and for this situation distances of 1.5–3.6 m are common and the level of the A-weighted background noise should be less than 50–60 dB. At public meetings or outdoors in yards, parks, or playgrounds distances between talker and listener are often from 3.7–9 m and the A-weighted-sound level of the background noise must be kept below 45–55 dB if nearly normal speech communication is to be possible.

3.4   LOUDNESS

The subjective loudness of sound refers to the magnitude of the auditory sensation produced in the person exposed to the sound. Loudness is a function of the intensity, frequency, bandwidth, and duration of the sound. Thus two sounds judged to be equally loud may substantially differ from each other. As indicated on p. 30 the unit of loudness level is the phon. Two problems exist with the loudness level scale measured in phons. First, equal increases in the judged loudness of a sound do not correspond to equal increases in the loudness level. For example, if the loudness is increased from 40 to 50 phons the judged loudness increases by a factor of two, whereas an increase in the loudness level from 50 to 70 phons yields a judged increase in loudness of a factor of four. Second, the sum of the loudness levels of two tones does not correspond to the judged loudness of the combined tones. If two tones of different frequency each have a loudness level of 60 phons, the judged loudness level of the combined tones would be 70 phons rather than 120 phons. For these reasons a subjective loudness scale in sones was introduced. The unit sone is defined as the loudness of a 1000 Hz tone with a sound pressure level of 40 dB re 20 μPa. Therefore, for a 1000 Hz tone a loudness of 1 sone corresponds to a loudness level of 40 phons. It should be emphasized that loudness levels are measured in phons and loudness is measured in sones. The relation between the loudness L in sones and the loudness level LL in phons is

$$L = 2^{(LL-40)/10} \qquad \text{sones} \tag{3-2}$$

or

$$\log_{10} L = 0.03 \, LL - 1.2$$

Conversely,

$$LL = 33.3 \log_{10} L + 40 \qquad \text{phons} \tag{3-3}$$

[The numerical relationship between L and LL as given by (3-2) and (3-3) is illustrated in the right hand ordinate of Figure 3-6.] In other words, loudness is found to increase much faster than loudness level, and each increase of 10 phons above 40 phons constitutes a doubling of loudness. Thus the sone scale has the advantage that a sound judged to be twice as loud as another one also has twice the numerical value of loudness.

3.5 CRITICAL BANDWIDTHS*

The range of audible frequencies is divided into 24 frequency intervals (called "frequency groups" or "critical bands") that in the midrange of pitch are about a third of an octave wide and are established by the characteristics of the ear itself. (Pitch is the subjective judgment of the frequency of a sound.) This division corresponds to a division of the basilar membrane into 24 approximately equally wide and 1.3-mm-long sections, which to some degree act as separate receptor units. The frequency groups are numbered progressively and are designated by their number. These critical bandwidths have been given the name bark in honor of H. Barkhausen, originator of the first loudness level unit.

The significance of the critical bandwidths is seen in the next section.

3.6 MASKING†

If the ear is exposed to two sounds at the same time, it perceives, under certain circumstances, only the louder one; the soft sound is masked by the loud one. The masked threshold is the sound pressure of the masked sound at which it becomes just audible. The masked threshold depends on the sound level of the masking sound and the spectra of the masked and the masking sounds.

As an example consider the masking of a tone by a noise of variable bandwidth. Figure 3-5 shows the masked threshold of the test tone as a function of its frequency. The masking noise is separated from a noise having frequency-independent spectral energy distribution (white noise) by means of filters having high selectivity; the sound intensity density is 40 dB re 10^{-12} W/m^2/Hz. The bandwidths are 16 kHz, 3175 Hz, 1.5 kHz, 160 Hz, and 16 Hz with a center frequency of 1 kHz (geometrical mean). The

---

*Reprinted, by permission, from *Physical and Applied Acoustics* by Meyer and Neumann, Academic Press, New York, 1972.
†*Ibid.*

lowest dashed curve in Figure 3-5 is the absolute threshold of hearing without masking noise. With the greatest noise bandwidth (16 kHz), the threshold of hearing for the test tone is raised considerably—by about 50 dB at 1000 Hz, for example.

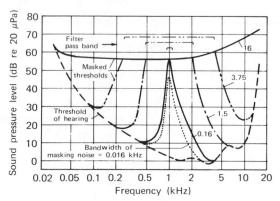

FIGURE 3-5
Masking threshold of a tone as a function of its frequency for masking by noise with an intensity density of $10^{-8}\,W/m^2/Hz$. Center frequency of the noise is 1000 Hz. (Reprinted, by permission, from *Physical and Applied Acoustics* by Meyer and Neumann, Academic Press, New York, 1972.)

With smaller bandwidths, the masking threshold within the bandwidth is the same as for broad-band noise; outside the limits of the band, it drops back to the absolute threshold of hearing. Since the decrease is always steeper at the lower limit of the band than at the upper limit, low-frequency noises mask higher frequencies better than high-frequency noises mask lower ones. Spectral components lying far outside the noise band are not masked. If the noise bandwidth is reduced, the sides of the masking threshold curve move together; the inflection points of the curves agree well with the cutoff frequencies of the bandpass filter and the height of the masking threshold inside the pass band remains unchanged. This is true only down to a critical bandwidth (160 Hz for 1000-Hz center frequency) at which point the trapezoidal form changes into a triangular one. If the bandwidth is made smaller than the critical value, the masking threshold falls but the triangular shape undergoes no further change (dotted curve for a bandwidth of 16 Hz).

The masking effect can be understood from the excitation characteristics of the basilar membrane. A masking tone sets the basilar membrane into vibration with maximal vibration not only at the particular point that corresponds to the exciting frequency, but also in a wider region, especially in the direction of the windows. It must then be assumed that another tone only slightly different in frequency will not be perceived until it produces a displacement of the basilar membrane at the point corresponding to its frequency that is stronger than the displacement produced by the first signal. To investigate the masking threshold as a function of frequency (e.g., Figure 3-5), the excitation of the basilar membrane by the masking sound is sampled with the test tone. Important facts concerning the charactersitics of the ear are obtained in this way.

Hence the bandwidths that can be measured by a person's ability to detect a pure tone in the presence of white noise are called critical bandwidths. Furthermore, the level of the pure tone must exceed slightly the total rms level of the noise in a critical bandwidth if the tone is to be heard. Another way of saying it is that the masking effect of the noise increases in proportion to its bandwidth until the critical bandwidth is reached. Beyond that bandwidth no further masking of the pure tone occurs. Still another way to interpret the critical bandwidth is with regard to loudness. Holding the sound-pressure level of the noise without the tone constant and decreasing the bandwidth below the critical value will not change the subjective loudness. On the other hand, again maintaining a constant sound-pressure level of the noise without the tone, but increasing the bandwidth beyond the critical level, will produce an increase in loudness.

3.7 STEVENS AND ZWICKER METHODS TO CALCULATE LOUDNESS

Introduction

Both Stevens and Zwicker have developed schemes for calculating the loudness of a complex noise, as a substitute for an auditory comparison. Indeed, so rarely is the auditory comparison actually made nowadays that the term "loudness level" almost automatically implies a calculation procedure. Both methods take into account the masking of one component of a complex noise by the other components. Both procedures have been standardized.

Stevens Mark VI[5]

Stevens' Mark VI method assumes a diffuse* sound field and takes as

---

*A field is diffuse if a great many reflected sound waves cross a given point in space from all possible directions such that the sound energy is uniform through the field around the point.

the starting point a frequency band analysis (either octave, half-octave, or 1/3-octave) of the noise to be evaluated. For each of the frequency bands, a loudness index is determined from a set of contours of equal loudness index as a function of band sound-pressure level and frequency, represent- ing the potential contribution of that band to the total loudness. These contributions are added together, each with a weighting factor, which is unity for the band with the largest loudness index, and which is assigned a value, F, for all other bands, depending on the analysis bandwidth as follows: for a 1/3-octave bandwidth $F = 0.15$; for a 1/2-octave bandwidth $F = 0.20$; for a 1/1-octave bandwidth $F = 0.3$.

The weighting factor, F, is intended to account for the fact that the loudest band inhibits, or masks, the contributions of the other bands to the total impression of the sound. This procedure implies that the masking caused by the loudest band is symmetrical, that is, that it inhibits equally the contributions of bands above and below it in frequency.

Stevens constructed a series of contours that are straight line approxi- mations of the equal loudness level contours. These curves are called loudness index contours and are given as a function of band pressure level and frequency in Figure 3-6. The loudness index contours all have a slope of -3 dB/octave, excepting that the slope is $+12$ dB/octave above 9000 Hz and -6 dB/octave below a certain frequency. The frequency at which the slope changes from -3 dB/octave to -6 dB/octave changes as a function of the loudness index. The frequency of this change in slope lies on a line having a slope of -21 dB/octave that passes through the point defined by a 10 dB band-pressure level and 1000 Hz.

Stevens' procedure for calculating loudness is summarized as follows:

1. Determine from the loudness index contours in Figure 3-6 the loudness index for each octave or partial (1/2- or 1/3-) octave bandwidth in terms of the center frequency of the band and the band pressure level.
2. Determine the total loudness (S_t) in sones from the formula

$$S_t = S_m (1 - F) + F \sum_{j=1}^{N} S_j \qquad \text{sones} \qquad (3\text{-}4)$$

    where S_m is the numerically largest loudness index in the data and $\sum_{j=1}^{N} S_j$ is the sum of all the loudness indexes over the band (including $j = m$), and F is a bandwidth correction factor given above.
3. The total loudness in sones may be converted into loudness level in phons by (3-3) or by the nomograph on the right-hand side of Figure 3-6.

Example 3-2. From the 1/3-octave band data given in Table 3-3 deter- mine the loudness level in phons by the Stevens method.

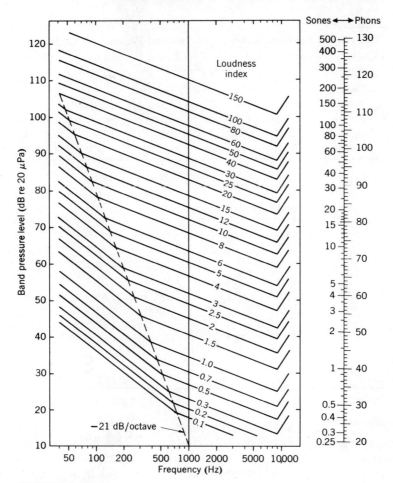

FIGURE 3-6
Contours of equal loudness index. (From Reference 5.)

Solution: The loudness index is determined from Figure 3-6 and is presented in Table 3-3. Using (3-4) with $F = 0.15$ and the results in Table 3-3 we have

$$S_t = 17(1 - 0.15) + 0.15(297.7) = 59.1 \text{ sones}$$

From (3-3) (or the right-hand side of Figure 3-6)

$$LL = 33.3 \log_{10}(59.1) + 40 = 99 \text{ phons}$$

TABLE 3-3
Tabulation of Loudness Index for Given Band-Pressure Levels in 1/3-Octave Bands

1/3-Octave band center frequency (Hz)	Band pressure level (dB)	Loudness index (sones)	1/3-Octave band center frequency (Hz)	Band pressure level (dB)	Loudness index (sones)
50	87.5	10	800	80.5	16
63	86	9.5	1000	76.7	13.2
80	83	10	1250	77	14
100	83	11	1600	75.5	13.5
125	81	10	2000	72	11.5
160	80	9.8	2500	70.5	11.5
200	84.7	14	3150	69.7	11.5
250	83.5	14.5	4000	68.3	11.5
315	79.5	11.5	5000	68.6	12.5
400	79.5	12.2	6300	67.5	12.5
500	81	15	8000	67.8	13.5
630	82.2	17	10000	68.1	12.0

$$\sum_{j=1}^{24} S_j = 297.7$$

Zwicker

Zwicker's latest procedure can accommodate both diffuse and free-field conditions but must begin from a third-octave band analysis of the noise. The calculations are considerably more complicated than Stevens' Mark VI because it takes into account the well-known asymmetry in masking, whereby an intense band of noise inhibits the contribution to loudness of bands higher in frequency than itself to a much greater degree than of bands lower in the frequency scale. The calculation is graphical by nature and involves planimeter measurement of a constructed area on a graph of band level against a distorted frequency scale; the analytic equivalent of this procedure is so complex that a computer is necessary if more than a few calculations are to be made. The method is, however, based more firmly on theory than the simpler Stevens approach and will work for more irregular spectra. Because the Zwicker method takes into account the strong upward spread of masking and Stevens' method does not, the results of Zwicker's calculations are typically 5 dB greater than Stevens' values for the same noises.

Because of the complexity of the Zwicker method the details will not be given here. However, complete examples are given in references 6 and 7.

3.8 PERCEIVED NOISE LEVEL (PNL)[8,9]

Studies of perceived noisiness have revealed that high-frequency noise is judged to be noisier than equally loud noise of lower frequency. Noise whose intensity varies radically with time is judged noisier than the same sound when the intensity remains relatively constant in time. Noise whose location is not observable is judged noisier than the same noise when it can be localized. Judged noisiness of sound varies inversely with the rise time of the sound; that is, if two sounds are increased to the same ultimate sound-pressure level the one with the fastest increase up to the maximum sound-pressure level is judged noisier. If one of two equally intense sounds contains tones or energy concentrated in narrow band-widths it will be judged noisier.

The first step in quantifying perceived noisiness is to construct curves of equal noisiness. These contours are plotted as a function of band sound-pressure levels and frequency in Figure 3-7 and resemble equal loudness contours. Since perceived noisiness is a function of the sound-pressure level, the frequency spectrum, and the temporal behavior of the noise, the reference test noise used in constructing equal perceived noisiness contours must be specified exactly. The recommended reference sound for perceived noisiness comparisons is an octave band of noise centered at 1000 Hz with the following temporal characteristics: the sound-pressure level of the noise increases in magnitude at a rate of 5 dB/ sec to its maximum value, which is maintained for an interval of 2 sec, and then the noise decreases at a rate of 5 dB/ sec.

The unit for perceived noisiness is the noy. The noy scale, like the sone scale for loudness, is chosen so that the unit is directly proportional to the sensation of perceived noisiness. Thus a sound whose perceived noisiness is 3 noys is judged to be three times as noisy as a sound of 1 noy. One noy is the judged noisiness of an octave band of noise centered at 1000 Hz with a sound pressure level of 40 dB and with the temporal characteristics described above. Perceived noisiness in noys can be converted to a dB-like scale. The quantity is then called the perceived noise level (PNL) and it is expressed in PNdB. A doubling of the perceived noisiness in noys is equivalent to an increase of 10 PNdB in the PNL. It is customary to report the PNL in PNdB rather than the perceived noisiness in noys.

The procedure for calculating the PNL is similar to Stevens' method for calculating loudness. The total perceived noisiness PN in noys is determined by adding the perceived noisiness of each frequency band interval

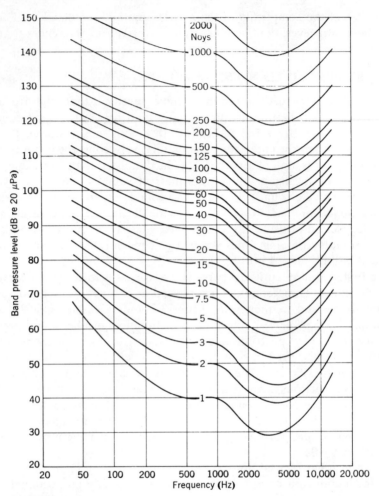

FIGURE 3-7
Contours of perceived noisiness. (From Reference 8.)

using the equation:

$$PN = N_{max}(1 - F) + F \sum_{j=1}^{M} N_j \qquad \text{noys} \qquad (3\text{-}5)$$

where N_{max} is the maximum value of N_j, $\sum_{j=1}^{M} N_j$ is the sum of the noisiness over all the frequency bands, and F is 0.3 for octave bands of noise or 0.15 for 1/3-octave bands of noise. The noy values for each frequency interval are determined from Figure 3-7. The perceived noise level, PNL in PNdB, is determined from the perceived noisiness, PN, by

$$PN = 2^{(PNL-40)/10} \qquad \text{noys} \qquad (3\text{-}6)$$

or

$$PNL = 33.3 \log_{10}(PN) + 40 \qquad PNdB \qquad (3\text{-}7)$$

Note the similarity of (3-5), (3-6), and (3-7) with (3-4), (3-2), and (3-3), respectively.

Example 3-3. Using the 1/3-octave data given in Table 3-3, determine the perceived noise level in PNdB.

Solution: The results are tabulated in Table 3-4. From this table it is seen $N_{max} = 18.7$. Using (3-5) and Table 3-4 yields

$$PN = 18.7(1 - 0.15) + 0.15(336.3) = 66.34 \text{ noys}$$

From (3-7) the PNL is obtained as follows

$$PNL = 33.3 \log_{10}(66.34) + 40 = 100.7 \text{ PNdB}$$

TABLE 3-4
Tabulation of Perceived Noise for Given 1/3-Octave Band Pressure Levels

1/3-Octave band center frequency (Hz)	Band pressure level (dB)	Perceived noisiness (noys)	1/3-Octave band center frequency (Hz)	Band pressure level (dB)	Perceived noisiness (noys)
50	87.5	10.6	800	80.2	16.2
63	86	11.3	1000	76.7	12.7
80	83	9.8	1250	77	14.9
100	83	12.1	1600	75.5	17.6
125	81	11.3	2000	72	15.8
160	80	11.3	2500	70.5	14.4
200	84.7	18.0	3150	69.7	16.6
250	83.5	17.8	4000	68.3	15.0
315	79.5	14.1	5000	68.6	14.3
400	79.5	15.4	6300	67.5	12.4
500	81	17.1	8000	67.8	10.3
630	82.2	18.7	10000	68.1	8.6
				$\sum\limits_{j=1}^{24} N_j = 336.3$		

3.9 NOISE CRITERIA (NC) CURVES

Noise criteria (NC) curves, introduced by Beranek, have been widely accepted in the United States and have come into extensive use, both as a means of evaluating existing noise problems and also to define design goals for achieving noise backgrounds that will be judged acceptable by the occupants. The method is based on both the Speech Interference Level (SIL) and the Loudness Level (LL). Nevertheless, if a background noise whose spectrum conforms to an NC curve shape is deliberately generated, it is discovered that it is not a pleasant or neutral noise, but sounds both "hissy" and "rumbly". This is ordinarily no disadvantage to using the NC curves for rating an existing noise; but it can lead to trouble as a design goal if, for example, the noise control efforts should actually yield a background noise matching an NC curve. According to current standards of acoustical comfort, such a spectrum would not be pleasing. A recent tendency in consulting practice, therefore, has been to adopt as a matter of course more stringent criteria than Beranek recommended from his studies.

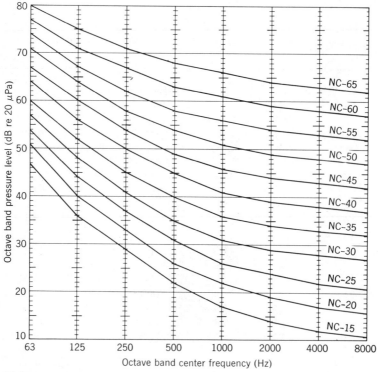

FIGURE 3-8
Noise criteria (NC) curves.

The NC curves are given in Figure 3-8. The NC values apply to steady noises and specify the maximum noise levels permitted in each octave band for a specified NC curve. For example, if an architectural specification calls for noise levels not to exceed the NC-20 criterion curve, the sound-pressure levels in all eight octave bands must be less than those of the NC-20 curve of Figure 3-8. Alternatively, the "NC rating" of a given noise equals the highest penetration of any band of that noise into the curves. For example, if a noise spectrum had octave-band levels, starting with the 63-Hz band, of 50, 55, 58, 60, 55, 45, and 39 dB, respectively, the noise would be given a rating of NC-57, because the fourth (500 Hz) band penetrates to a level that is 2 dB above the highest NC curve approached, that is, the NC-55 curve.

A similar procedure has been under consideration in the International Organization for Standardization since 1961; although none of the draft documents describing this rating has been accepted yet as an ISO Recommendation, the method has appeared in the literature and is frequently mentioned in the context of building noise control. The ISO curves are called either N or NR (for Noise Rating) curves, and the ISO proposals grew directly out of the work of Beranek. Table 3-5 gives recommended NC values for different environments.

To overcome the objections cited above to the NC curves Beranek et al[10] introduced the preferred noise criteria (PNC) curves shown in Figure 3-9. They are used in exactly the same manner as the old NC curves. Table 3-5 gives recommended PNC values for different room environments.

TABLE 3-5
Recommended Category Classification and Suggested Noise Criteria Range for Steady Background Noise as Heard in Various Indoor Functional Activity Areas[a]

Type of space (and acoustical requirements)	PNC Curve	NC Curve
Concert halls, opera houses, and recital halls (for listening to faint musical sounds).	10–20	10–20
Broadcast and recording studios (distant microphone pick-up used).	10–20	15–20
Large auditoriums, large drama theaters, and churches (for excellent listening conditions).	Not to exceed 20	20–25
Broadcast, television, and recording studios (close microphone pick-up only).	Not to exceed 25	20–25

TABLE 3-5 (Continued)

Type of space (and acoustical requirements)	PNC Curve	NC Curve
Small auditoriums, small theaters, small churches, music rehearsal rooms, large meeting and conference rooms (for good listening), or executive offices and conference rooms for 50 people (no amplification).	Not to exceed 35	25–30
Bedrooms, sleeping quarters, hospitals, residences, apartments, hotels, motels, and so forth (for sleeping, resting, relaxing).	25–40	25–35
Private or semiprivate offices, small conference rooms, classrooms, libraries, and so forth (for good listening conditions).	30–40	30–35
Living rooms and similar spaces in dwellings (for conversing or listening to radio and TV).	30–40	35–45
Large offices, reception areas, retail shops and stores, cafeterias, restaurants, and so forth (for moderately good listening conditions).	35–45	35–50
Lobbies, laboratory work spaces, drafting and engineering rooms, general secretarial areas (for fair listening conditions).	40–50	40–45
Light maintenance shops, office and computer equipment rooms, kitches, and laundries (for moderately fair listening conditions).	45–55	45–60
Shops, garages, power-plant control rooms, and so forth (for just acceptable speech and telephone communication). Levels above PNC-60 are not recommended for any office or communication situation.	50–60	—
For work spaces where speech or telephone communication is not required, but where there must be no risk of hearing damage.	60–75	—

[a]After Reference 10.

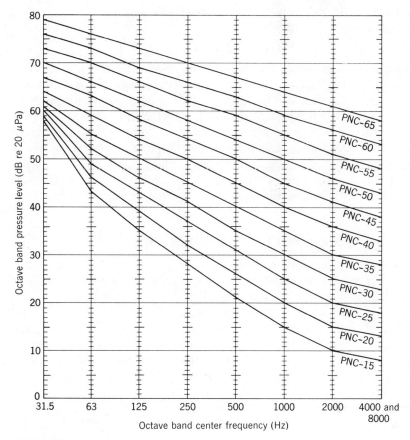

FIGURE 3-9
Preferred noise criteria (PNC) curves.

3.10 TRAFFIC NOISE INDEX (TNI)

The base measure for the traffice noise index (TNI) is the A-weighted-sound level sampled at numerous discrete intervals outdoors over a 24-hr period. From the statistics of these sampled levels two sound levels, denoted L_{10} and L_{90}, are determined. The L_{10} level is the A-weighted level that is exceeded 10% of the time and the L_{90} is the A-weighted level exceeded 90% of the time. Thus the L_{10} level is an indication of the peak levels of the intruding noise, whereas the L_{90} level is an indicator of the background level into which the L_{10} levels intrude. In Section 3.11 the procedure used to determine these levels is discussed in detail.

The Traffic Noise Index, a weighted combination of L_{10} and L_{90}, is defined as:

$$TNI = 4(L_{10} - L_{90}) + L_{90} - 30 \qquad (3\text{-}8)$$

The first term expresses the range of the "noise climate" and describes the variability of the noise and the second represents the background noise level; the third term is introduced to yield more convenient numbers.

The real step forward made by the TNI is that it breaks away from other, earlier ratings of noise exposure to emphasize that a significant degree of annoyance arises from the variable character of the noise, represented by the difference between L_{10} and L_{90}. This has not appeared in any form in the ratings discussed so far and has been a severe handicap in dealing with as variable a stimulus as urban traffic noise.

3.11 NOISE POLLUTION LEVEL (NPL)

Before proceeding with the definition of the noise pollution level it is necessary to introduce and define several statistical quantities, namely, L_{10}, L_{50}, L_{90}, and a quantity denoted L_{eq}. Consider the portion of a time-varying record of the A-weighted-sound level shown in Figure 3-10. At each uniformly spaced interval of time t_k, $k = 1, 2, \ldots$, the corresponding A-weighted level L_k'' is obtained. Let the total range of A-weighted levels be divided into N equal levels, such that L_1' is the minimum level considered, L_{N+1}' is the maximum level, and $L_{j+1}' = L_j' + \Delta L$, $j = 1, 2, \ldots N$, where

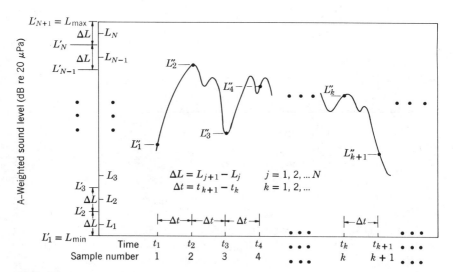

FIGURE 3-10
Quantization of time-varying A-weighted-sound levels sampled at discrete intervals.

$\Delta L = (L'_{N+1} - L'_1)/N$. The number of L''_k that fall within the interval $L'_j \leqslant L''_k < L'_{j+1}$, is denoted M_j, $j = 1, 2, \ldots N$, $k = 1, 2, \ldots$. The total number of samples accumulated is $M = \sum_{j=1}^{N} M_j$. Table 3-6 is now constructed as a function of the class interval L_j, where $L_j = (L'_j + L'_{j+1})/2$, $j = 1, 2, \ldots N$. From the third column of Table 3-6 one obtains a histogram of the amplitude distribution of the L_j as shown in Figure 3-11a. The fourth column yields the cumulative probability from which the L_{10}, L_{50}, and L_{90} levels can be determined. A typical curve is shown in Figure 3-11b.

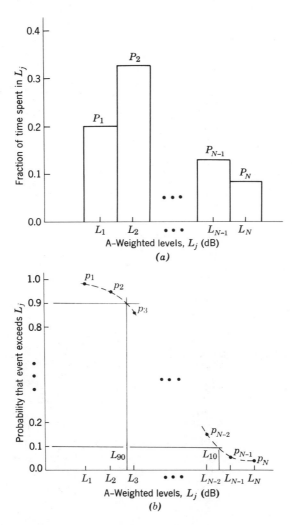

FIGURE 3-11
(*a*) **Histogram resulting from data obtained from the third column of Table 3-6.**
(*b*) **Cumulative distribution function obtained from the fourth column of Table 3-6.**

TABLE 3-6
Determination of Histogram and Cumulative Probability from a Sampled Record of the A-Weighted-Sound Level Versus Time

Class interval (L_j)	Number of events in L_j (M_j)	Fraction of time spent in L_j (P_j)	Probability that level exceeds L_j (p_j)
L_1	M_1	$P_1 = \dfrac{M_1}{M}$	$p_1 = \sum\limits_{j=1}^{N} P_j = 1$
L_2	M_2	$P_2 = \dfrac{M_2}{M}$	$p_2 = \sum\limits_{j=2}^{N} P_j = \dfrac{1}{M} \sum\limits_{j=2}^{N} M_j$
$\vdots$	$\vdots$	$\vdots$	$\vdots$
L_{N-1}	M_{N-1}	$P_{N-1} = \dfrac{M_{N-1}}{M}$	$p_{N-1} = \sum\limits_{j=N-1}^{N} P_j = \dfrac{1}{M} \sum\limits_{j=N-1}^{N} M_j$
L_N	M_N	$P_N = \dfrac{M_N}{M}$	$p_N = P_N = \dfrac{M_N}{M}$
	$M = \sum\limits_{j=1}^{N} M_j$	$\sum\limits_{j=1}^{N} P_j = 1$	

The results obtained from Table 3-6 are now used to determine a quantity called the "energy mean," L_{eq}, of the event described in Figure 3-10, as follows:

$$L_{eq} = 10 \log_{10} \left[\sum_{j=1}^{N} P_j \times 10^{(L_j/10)} \right] \qquad \text{dB re 20 } \mu\text{Pa} \qquad (3\text{-}9)$$

which has a standard deviation

$$\sigma = \left[\sum_{j=1}^{N} P_j L_j^2 - \left\{ \sum_{j=1}^{N} P_j L_j \right\}^2 \right]^{1/2} \quad \text{dB} \qquad (3\text{-}10)$$

We now proceed to the definition of the noise pollution level.

The noise pollution level (NPL) is based on two terms, one representing the equivalent continuous noise level and the other representing the annoyance due to the fluctuations of the noise level. It is determined from the expression

$$L_{NP} = L_{eq} + k\sigma \qquad (3\text{-}11)$$

where L_{eq} is the "energy mean" of the A-weighted-noise level over a specified period given by (3-9), σ is the standard deviation of the instantaneous level given by (3-10), and k is a constant tentatively set equal to 2.56, since this value leads to the best fit with currently available studies of subjective response to noise. The first term is determined largely by the intensity of the intruding noises (because of the logarithmic averaging), unless these occur so seldom that the background noise comprises most of the total noise exposure. The second term is determined by the time-dependence (specifically the variability in level) of the sequence of intruding noise events, rather than on the mean energy content, and is thus greatly influenced by the prevailing background noise; the lower the background noise, the greater the variability for a given sequence of intrusive events.

The period over which the noise pollution level is to be reckoned should be reasonably homogeneous with respect to both the noise occurrences and the activities of the population; one would, for example, distinguish between day and night. Not only the intrusive noise signals, but also the background noise, must be known (or assumed) in order to utilize the formula; the relevant background is the one actually experienced by the population in question, generally the indoor ambient noise. Thus if the intrusive noise levels are initially measured outdoors, they may be "brought indoors," via the attenuation of the exterior wall for incorporation into the time pattern of sound that the occupant experiences and which the noise pollution level evaluates. An equally valid procedure would be to project the indoor background levels outdoors.

For many community noise situations of interest one can use alternate (approximate) expressions for the noise pollution level as follows:

$$L_{\mathrm{NP}} = L_{eq} + L_{10} - L_{90} \qquad \mathrm{dB\ re\ 20\ \mu Pa} \qquad (3\text{-}12)$$

or

$$L_{\mathrm{NP}} = L_{50} + d + \frac{d^2}{60} \qquad \mathrm{dB\ re\ 20\ \mu Pa} \qquad (3\text{-}13)$$

where $d = (L_{10} - L_{90})$ and L_{10} and L_{90} are the decile A-weighted-sound-pressure levels exceeded 10 and 90 percent, respectively, of the time during the observation period, and L_{50} is the median level. From (3-12) and (3-13) it is seen that $L_{eq} \approx L_{50} + d^2/60$.

It should be pointed out that the NPL was developed in such a way that it accounts reasonably well for the results of several previous studies in which subjective response to aircraft and road traffic noise was compared with physical measures of the noise exposure. However, there is not yet any data from subsequenct tests performed specifically to determine how well the noise pollution level correlates with observed subjective response to these, or to any other kinds of noise: real, recorded, or simulated. Nor have any studies been conducted in which the NPL is included in comparison tests to assess the merits of different rating procedures for predicting some particular aspect of human response.

The NPL is used by the U. S. Department of Housing and Urban Development (HUD) as a guide either to zoning or to the siting and construction of dwellings where the noise situation and the zoning are already established. In general, the *indoor* A-weighted-noise levels preferably should not exceed those listed below.

1. Listening to radio and television:
    $L_{50} = 35-45$ dB
    $L_{10} = 41-51$ dB
    $L_{NP} = 50-60$ dB
2. Sleeping:
    $L_{50} = 25-50$ dB
    $L_{10} = 31-56$ dB
    $L_{NP} = 40-65$ dB

The range of values given for sleeping depends on the nature of the surroundings, for example country versus urban areas. To convert the above figures to approximate *outdoor* levels, add to each number 10 dB for open windows and 20 dB for closed windows with openable sashes.

HUD further sets external noise exposure standards for new residential construction as follows:

	Approximate NPL
Clearly unacceptable	>88
Exceeds an A-weighted level of 80 dB for 60 min/24 hr	
Exceeds an A-weighted level of 75 dB for 8 hr/24 hr	
Normally unacceptable	74–88
Exceeds an A-weighted level of 65 dB for 8 hr/24 hr	
Loud repetitive sounds on site	
Normally acceptable	62–74
Does not exceed an A-weighted level of 65 dB for more than 8 hr/24 hr	
Clearly acceptable	<62
Does not exceed an A-weighted level of 45 dB for more than 30 min/24 hr	

There are two ways the environment can be sampled to obtain the data necessary to use (3-9) and (3-10). One method is called "continuous" sampling and the other "manual" sampling. Fisk[11] has shown from theoretical considerations that a good approximation to continuous sampling is obtained with 2 samples/sec. A faster sampling rate reduces the statistical error, but the improvement is only 1% better. Using at least the 2 samples/sec sampling rate he has shown that the sampling error ΔL_{eq} can be estimated from the expression

$$\Delta L_{eq} \sim \frac{7.4}{\sqrt{m}} \text{ dB}$$

where m is the number of vehicles passing during the total sampling period. Consequently light traffic (<300 vehicles/hr) requires the total sampling period to be relatively long.

The manual techniques uses only a top watch and a sound level meter on fast response (see p. 110). For this procedure the sound level meter is read every 10 sec. (This sampling interval can be slightly different but not shorter than 5 sec.) The details of this technique are not discussed here; however, they are given in great detail by Yerges and Bollinger[12] and in a Federal Highway Administration monograph.[13] The latter's approach is restricted to the determination of the L_{10} level only, whereas the former is somewhat more general in that L_{10}, L_{50}, and L_{90} can each be obtained with the same statistical confidence limits. Although both procedures appear different they are mathematically equivalent.

Example 3-4. From the data presented in Table 3-7, which was obtained from the statistical analysis of noise at an airport, determine the noise pollution level. These tabular values are also shown in Figure 3-12.

TABLE 3-7
Statistics of Noise at an Airport

A-Weighted interval $(L'_{j+1} - L'_j)$ (dB)	A-Weighted level of center of interval, L_j (dB)	Percentage of time spent in interval, L_j ($P_j \times 100$)	Percentage of time A-weighted level exceeded interval level ($p_j \times 100$)
75.1– 77.5	76.3	0.011	100.00
77.6– 80.0	78.8	0.007	99.99
80.1– 82.5	81.3	5.58	99.98
82.6– 85.0	83.8	25.1	94.4
85.1– 87.5	86.3	26.6	69.3
87.6– 90.0	88.8	21.2	42.7
90.1– 92.5	91.3	8.33	21.5
92.6– 95.0	93.8	9.45	13.2
95.1– 97.5	96.3	3.02	3.72
97.6–100.0	98.8	0.696	0.696

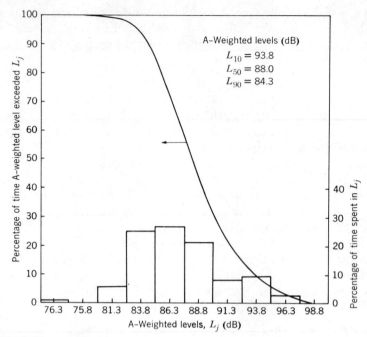

FIGURE 3-12
Histogram and cumulative distribution function for data shown in Table 3-7.

Solution: The "energy mean" is obtained from (3-9), yielding

$$L_{eq} = 10\log_{10}(0.011 \times 10^{7.63} + 0.007 \times 10^{7.88} + 0.0558 \times 10^{8.13} + 0.251 \times 10^{8.38}$$

$$+ 0.266 \times 10^{8.63} + 0.212 \times 10^{8.88} + 0.0833 \times 10^{9.13}$$

$$+ 0.0945 \times 10^{9.38} + 0.0302 \times 10^{9.63} + 0.00696 \times 10^{9.88})$$

$$L_{eq} = 10\log_{10}(8.635 \times 10^{8}) = 89.36 \text{ dB re } 20 \text{ } \mu\text{Pa}$$

The standard deviation is given by (3-10). Thus

$$\sigma = (7659.1476 - 7644.1705)^{1/2} = 3.87 \text{ dB}$$

Using (3-11) yields

$$L_{NP} = 89.36 + (2.56)(3.87) = 99.27 \text{ dB}$$

It is instructive to compare this value with that obtained using the approximate formulas (3-12) and (3-13). From Figure 3-12 we have that the A-weighted levels are $L_{90} = 84.3$ dB and $L_{10} = 93.8$ dB. From (3-12) we have

$$L_{NP} = 89.36 + 93.8 - 84.3 = 98.86 \text{ dB}$$

which is in agreement with the previous (exact) value. Now using (3-13) we find, since the A-weighted level $L_{50} = 88.0$ dB,

$$L_{NP} = 88.0 + 93.8 - 84.3 + \frac{(93.8 - 84.3)^2}{60} = 99.00 \text{ dB}$$

which is also in agreement with the original result.

3.12   NOISE EXPOSURE FORECAST (NEF)

Tone-Corrected Perceived Noise Level (PNLT)

The tone-corrected PNL is an adjustment applied to the PNL to increase its value when tones are present in the signal. One first calculates the PNL as shown in Section 3.8 and then adds the tone correction based on its frequency and the amount that the tone's amplitude exceeds the noise in the adjacent 1/3-octave bands. The procedure is somewhat involved and is not discussed here. However, a detailed explanation and a worked example can be found in Reference 6. The correction is typically on the order of 1–2 PNdB.

Effective Perceived Noise Level (EPNL)

The EPNL is a refinement of the PNLT and takes into account the signal duration relative to the maximum value of PNLT during the event. It is used by the Federal Aviation Administration in aircraft certification. The procedure to determine the EPNL is as follows: At each half-second interval of time from the start of the event (usually an aircraft flyover) the PNLT is determined as indicated above. Denote the value obtained during the half-second starting at t_j as $(PNLT)_j$, $j = 1, 2 \ldots J$. From the J $(PNLT)_j$ determine the maximum value and denote it PNLTM. For times earlier than the occurrence of PNLTM find the first value of j for which $(PNLT)_j \leqslant PNLTM - 10$. Call this value $(PNLT)_K$. For times later than the occurrence of PNLTM again find the first value of j for which $(PNLT)_j \leqslant PNLTM - 10$. Call this value $(PNLT)_M$. The Effective Perceived Noise

Level can now be defined as

$$\text{EPNL} = 10 \log_{10} \sum_{j=K}^{M} 10^{[(\text{PNLT})_j/10]} - 13 \qquad \text{EPNdB} \qquad (3\text{-}14)$$

NEF

The NEF technique is an extension of an older technique called the Composite Noise Rating (CNR); it applies only to commercial aircraft. The total noise exposure at a given point is composed of noise produced by different aircraft flying different flight paths. For a specific class of aircraft i, on flight path j, the NEF_{ij} can be expressed:

$$\text{NEF}_{ij} = \text{EPNL}_{ij} + 10 \log_{10} \left(n_{D_{ij}} + 16.67 n_{N_{ij}} \right) - 88 \qquad \text{EPNdB} \quad (3\text{-}15)$$

where $n_{D_{ij}}$ and $n_{N_{ij}}$ are the numbers of operations for daytime (0700–2200 hr) and nighttime (2200–0700 hr), respectively, of aircraft class i, on flight path j. The choice of these constants signifies that a single nighttime flight contributes as much to the NEF as 17 daytime flights; the number 88 is arbitrarily chosen so that NEF numbers typically lie in a range where they are not likely to be confused with other composite noise ratings.

The total NEF at a given gound position is determined by adding all the individual NEF_{ij} values on an energy basis:

$$\text{NEF} = 10 \log_{10} \sum_{i} \sum_{j} 10^{(\text{NEF}_{ij}/10)} \qquad (3\text{-}16)$$

The noise exposure forecast technique does predict community response, as it purports to do, but its predictions are rough. In its present formulation, it is unable to account for variations between stimulus and response arising from differences between individuals or communities rather than from differences in noise-generating parameters. However, the technique does identify a point at which noise changes the environment such that few activities can be carried on without substantial alterations, many people complain, and people organize to combat aircraft noise. Essentially, the status of the NEF technique is that of a planning tool. With it a point of "noise saturation" can be defined and lower grades of "noise contamination" estimated.

The ranges of the numerical values of the NEF and their significance are given in Table 3-8.

TABLE 3-8
Land Use Compatibility Chart as a Function of NEF[a]

NEF	Residential	Commercial, industrial	Hotels, motels, offices, public buildings	Schools hospitals, churches	Theaters, auditoriums	Outdoor theaters
<24	Yes	Yes	Yes	Yes	(A)	(A)
24–30	Yes	Yes	Yes	(C)	(C)	No
30–40	(B)	Yes	(C)	No	No	No
>40	No	(C)	No	No	No	No

[a](A)—A detailed noise analysis should be undertaken for all indoor or outdoor music auditoriums and all outdoor theaters. (B)—Case history experience indicates that individuals in private residences may complain, perhaps vigorously. Concerted group action is possible. New single dwelling construction should generally be avoided. For high-density dwellings (apartments) construction, (C) will apply. (C) —Avoid construction unless a detailed analysis of noise reduction requirements is made and needed noise control features are included in building design.

The actual implementation of the NEF method is extremely difficult to perform without the aid of a large high speed digital computer and a very large amount of data describing the noise characteristics of each type of aircraft. However, the results of a recent study by Yeowart[14] given in Figure 3-13 suggest the PNdB limits of each individual noise event as a

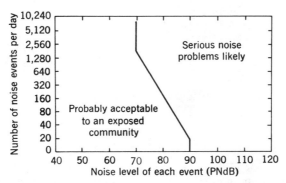

FIGURE 3-13
Acceptable aircraft noise levels to an exposed community as a function of the number of noise events. (Reprinted, by permission, from Reference 14.)

function of the number of noise events per day. As illustrated, as many as 20 events/day of 90 PNdB can occur without serious problems arising. However, as the number exceeds 1000 events/day (in the vicinity of a busy airport, for example) the level of each event must decrease to 70 PNdB.

3.13   DAY-NIGHT LEVEL (L_{dn})

The U. S. Environmental Protection Agency has introduced[3] the day-night level, L_{dn}, to provide a single number measure of community noise exposure over a specified period. It was designed to improve L_{eq} by adding a correction of 10 dB for the nighttime (2200–0700 hr) sound levels to account for the increased annoyance to noise during these hours. Its main purpose is to predict the effects on a population of the average long-term exposure to environmental noise. The L_{dn} is obtained from the relation

$$L_{dn} = 10\log_{10}[0.625 \times 10^{(L_d/10)} + 0.375 \times 10^{([L_n+10]/10)}] \quad\quad dB \quad (3\text{-}17)$$

where $L_d = L_{eq}$ for the daytime (0700–2200 hr) and $L_n = L_{eq}$ for the nighttime (2200–0700 hr).

The relation of typical numerical values of L_{dn} to various types of environments is as follows:

Environment	L_{dn} (dB)
Downtown major metropolis	75–80
Very noisy residential	68–73
Residential areas	
Noisy urban	63–68
Urban	58–63
Suburban	53–58
Small town and quiet suburban	46–53

For outdoor activities free of speech interference and annoyance an $L_{dn} \leqslant 55$ dB is recommended. For indoor residential areas a value of $L_{dn} \leqslant 45$ dB is recommended. If the 24-hr period used in the L_{dn} method is not applicable (e.g., schools, parks, etc.) the L_{eq} for that appropriate period is substituted for L_{dn}. However, the recommended levels remain the same.

3.14   COMPARISON OF THE RATINGS

The various rating schemes, based on physical measurements of a noise and described in this chapter, differ greatly in the amount of calculation effort required. The choice of a suitable rating for a specific purpose should weigh the trouble involved in determining the rating against the probable precision in predicting community reaction, as indicated in the interrating comparison studies described here.

Three different comparison approaches have been used: (1) comparing one noise rating with another, for a variety of different real or artificial noise spectra (with no appeal to human judgment) to see how well the ratings correlate with one another, or, more specifically, how well the more complicated ratings can be predicted from the simpler ones; (2) investigating which of the ratings assigns the most nearly uniform values to a series of noise events that are independently judged by people to be subjectively equivalent in some respect; or (3) determining the rating that most consistently puts each noise in the same qualitative category to which subjective judgements have assigned it.

The most significant of the interrating comparisons have been thoroughly reviewed by Schultz[1] and more recently by Ollerhead.[15] Omitting the details, we make use of some of Schultz's closing remarks and conclusions.

On the desirability, for every purpose, of the simpler ratings, such as the A-weighted sound level, as compared with the more complicated calculations of loudness level or perceived noise level we quote Botsford and Beranek, respectively.

> ...refinement of noise rating methods beyond sound levels is a futile exercise as far as improvement of ability to appraise human response of groups is concerned.

> The conclusion to be drawn from these various studies appears to be that, if a spectrum is (1) continuous in time, (2) continuous in frequency, (3) contains no sharp peaks or dips, and (4) extends over a wide frequency range, the A-weighted-sound level is as satisfactory a single-number method for rating noise as is the PNdB or loudness level in phons calculated by either the Zwicker or the Stevens method. In those instances, however, where a spectrum is quite non-uniform (some bands are much louder then others) or where it contains pure tones, or where the noise is intermittent, the A-weighted-sound level may not be a satisfactory measure of the subjective evaluation. Indeed, the other methods may not be entirely satisfactory, either.

Another expert in the field of aircraft noise is equally firm:

The best known measures of aircraft noise exposure, e.g., the NEF, are more highly correlated among themselves than is any one of them with annoyance, and they are about equally effective in conjunction with social variables in the prediction of annoyance. It follows that the elaborate procedures that have been devised for calculating the effective perceived noise level of aircraft flyovers offer no advantages over relatively simple measures in dealing with community reaction, although they are of great value in comparing aircraft and evaluating noise reduction.

It is difficult to select any one rating as best, or most accurate; that choice must be made in terms of the particular application in which it is to be used. The requirements for a suitable noise rating for estimating the magnitude of the transportation noise problem, for instance, will differ from those of a local planning board whose goal may be to establish local noise ordinances, or from those of an acoustical consultant who may need to design noise control measures for some specific noise source. A rating adequate to determine that one type of automobile or aircraft is more, or less, acceptable than another would perhaps differ from the rating adequate to certify that an individual household noise source meets a specified criterion. For each particular task one must select a tool that is neither more nor less complex than is needed for the job.

In view of the variety of situations where community noise ratings will be applied, it is premature to attempt to prescribe the most suitable rating for each application. Nevertheless, for any but the most specialized studies, it can be amply demonstrated that the A-weighted-sound level correlates as well as any of the ratings with subjective response and, in view of its simplicity, there is a great deal to recommend it for use in building codes, antinoise ordinances, automotive vehicle-type certification tests (i.e., measurements of the maximum noise-making capability of the vehicles), and monitoring of urban traffic noise.

There is increasing certainty that some account must be taken of the time-variability of urban noise if one is to predict adequately the human response. Both the Traffic Noise Index and the Noise Pollution Level are ratings designed to deal with this question. It is too early to decide which, if either, of these formulas will be best suited, in the long run, to the description of urban noise. No doubt the ultimate rating will bear a strong resemblance to these two ratings, since they account so well for a substantial amount of data.

REFERENCES

1. T. J. Schultz, *Community Noise Ratings*, Applied Science Publishers, London (1972), pp. 66–84.

2. J. Miller, "Effects of Noise on People," Report No. NTID 300.7, U. S. Environmental Protection Agency, Washington, D. C. (December 31, 1971).

3. "Information of Levels of Environmental Noise Requisite to Protect Public Health and Welfare with an Adequate Margin of Safety," Report No. 550/9-74-004, U. S. Environmental Protection Agency, Washington, D. C. (March 1974).

4. E. Meyer and E. Neumann, *Physical and Applied Acoustics*, Academic Press, N. Y. (1972), pp. 246–250.

5. "Procedure for the Computation of Loudness of Noise," American National Standard USAS S3.4-1968, American National Standards Institute, New York, N. Y. (1968).

6. K. S. Pearsons and R. L. Bennett, "Handbook of Noise Ratings," NASA CR-2376, National Aeronautics and Space Administration, Washington, D. C. (April 1974), pp. 32–49.

7. "Fundamentals of Noise: Measurement, Rating Schemes, and Standards," Report No. NTID 300.15, U. S. Environmental Protection Agency, Washington, D. C. (December 31, 1971), pp. 55–56.

8. K. D. Kryter, *The Effects of Noise on Man*, Academic Press, N. Y. (1970), Chapter 8.

9. K. D. Kryter and K. S. Pearsons, "Some Effects of Spectral Content and Duration on Perceived Noise Levels," *J. Acoust. Soc. Am.*, Vol. 35, No. 6 (June 1963), pp. 866–883.

10. L. L. Beranek, W. E. Blazier, and J. J. Figwer, "Preferred Noise Criterion (PNC) Curves and their Application to Rooms," *J. Acoust. Soc. Am.*, Vol. 50, No. 5 (November 1971), pp. 1223–1228.

11. D. J. Fisk, "Statistical Sampling in Community Noise Measurements," *J. Sound Vib.*, Vol. 30, No. 2 (1973) pp. 222–236.

12. J. F. Yerges and J. Bollinger, "Manual Traffic Noise Sampling—Can It Be Done Accurately?," *Sound and Vibration* (December 1973) pp. 23–30.

13. "Fundamentals and Abatement of Highway Traffic Noise," Report No. FHWA-HHI-HEV-73-7979-1, U. S. Department of Transportation, Federal Highway Administration, Washington, D. C. (June 1973), pp. 3–9 to 3–16.

14. N. S. Yeowart, "An Acceptable Exposure Level for Aircraft Noise in Residential Communities," *J. Sound Vib.*, Vol. 25, No. 2 (1972), pp. 245–254.

15. J. B. Ollerhead, "Scaling Aircraft Noise Perception," *J. Sound Vib.*, Vol. 26, No. 3 (1973), pp. 361–388.

4

INSTRUMENTATION
FOR THE
MEASUREMENT
AND ANALYSIS
OF SOUND*

4.1 ANALYSIS OF SIGNALS

Introduction

Observed signals representing physical phenomena can be classified as either deterministic or nondeterministic. Deterministic signals are those that can be described by an explicit mathematical relationship as a function of time. Nondeterministic signals cannot and are usually termed random.

Deterministic signals can be further classified as either periodic or aperiodic (transient). An example of a periodic signal is a single frequency

*For a more complete treatment of the subject see Reference 1.

tone played for a very long time. An example of an aperiodic signal would be a sonic boom. Nondeterministic signals also have several subclasses. However, for our purposes we consider only those that can be classified as ergodic, that is, those signals from which statistically meaningful information can be obtained from one relatively short sample. An example of this type of process is constant speed machinery noise. Traffic noise, on the other hand, is not an ergodic random process.

We now examine the types of information that can be obtained from the three types of signals: periodic, aperiodic, and random.

Periodic Signals

A periodic signal is one for which its shape (amplitude) as a function of time repeats itself in its entirety every time interval, or period, T_0. The frequency at which the signal is periodic is given by $f_0 = 1/T_0$ (Hz) which, we recall, is related to the circular frequency ω_0 by $\omega_0 = 2\pi f_0$ (rad/sec). It can be shown that every physically realizable periodic signal is composed of a suitably weighted combination of single frequency tones, with the frequency of each tone being some integer multiple of the fundamental frequency, f_0. If the amplitude of the nth tone, or harmonic, is C_n, then C_n corresponds to the tone having a frequency nf_0. We see that C_1 is the amplitude of the fundamental tone f_0.

A useful method for describing the content of a periodic signal, $g(t)$, is its power spectrum. The power spectrum is given by

$$P_{av} = \frac{1}{T} \int_0^T g^2(t) dt = \frac{1}{2} \sum_{n=1}^{\infty} C_n^2 = [\text{rms value of } g(t)]^2 \qquad (4\text{-}1)$$

wherein we have omitted the dc (zero frequency) term. The average power of the periodic signal excluding its dc contribution, if any, is designated P_{av}. A typical ideal power spectrum plot is shown in Figure 4-1. We see that the power spectrum plot for a periodic signal is discrete. The sum of the amplitude squared of each component gives the average power of the signal. The spacing between each harmonic, Δf, is proportional to $1/T_0$. As T_0 becomes larger, Δf becomes smaller and the spectral lines crowd together. If one were measuring sound-pressure level in units of μPa, then through a suitable calibration procedure discussed in Section 4.5, it can be shown that the ordinate of Figure 4-1 is proportional to $(\mu Pa)^2$.

Before leaving the discussion of periodic signals, it is worthwhile to introduce the definition of *percentage harmonic distortion*. In many instances one deals with single frequency tones. Oftentimes this single frequency tone passes through systems that introduce harmonics that were not present in the original signal. (This is essentially a nonlinear alteration

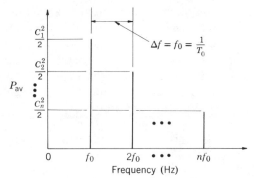

FIGURE 4-1
Typical power spectrum for a periodic signal.

of the signal.) A measure of how much these higher harmonics added to the original signal leads to the following definition of percentage harmonic distortion, % HD:

$$\% \text{ HD} = \frac{100}{|C_1|} \left(\sum_{n=2}^{\infty} C_n^2 \right)^{1/2} \% \qquad (4\text{-}2)$$

It is seen that the square of the numerator of (4-2) is simply the difference of the total average power P_{av} and the "power" of the fundamental (original) frequency component. For $\% \text{ HD} \leqslant 10\%$, $|C_1|$ can be replaced by $\sqrt{P_{av}}$ to a very good approximation.

Aperiodic Signals

An aperiodic signal may be considered as a periodic signal of infinite period. Returning to Figure 4-1 it is seen that as T_0 became very large, the discrete line spectrum crowds together. In fact, these lines can be packed as close together as desired by making T_0 large enough. When actually taken to the limit the discrete line spectrum of the periodic signal goes over to a continuous spectrum. Furthermore, the ordinate becomes the energy density, the amount of energy per Hz. A typical plot of the energy density spectrum, $S(f)$, is shown in Figure 4-2. The area under the curve in Figure 4-2 yields the total energy in the signal. If one were measuring sound pressure in units of μPa, the ordinate of Figure 4-2 would be proportional to $[(\mu\text{Pa})^2 - \text{sec}]/\text{Hz}$.

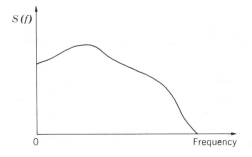

$S(f)$

0   Frequency

FIGURE 4-2
Typical plot of the energy spectral density for an aperiodic signal.

When trying to analyze aperiodic signals in a practical situation, one often resorts to a "tape loop". A tape loop simply plays back the aperiodic signal, at a fixed rate, over and over again. Thus the aperiodic signal is made periodic. The resulting analysis of the periodic signal yields a spectral plot like that given in Figure 4-1. To get the "correct" result, a line is drawn connecting the peaks of the C_n^2. To have enough spectral lines to trace out the loci of these peaks T/T_0 should have a value of 0.3–0.2, where T is the duration of the aperiodic signal and T_0 is the reciprocal of the repetition frequency (the period of the tape loop). The second requirement is to multiply the ordinate, which is the average power, by T_0^2.

Random Signals

When analyzing random signals, the average power is defined in a small band of frequencies, B, centered around f_c. In this small band of frequencies the signal is averaged over a time T. If T is sufficiently long and B is sufficiently narrow, the power spectral density $G(f_c)$ centered at f_c, is defined as

$$G(f_c) = \frac{P_{av}(B,f_c,T)}{B} \qquad (4\text{-}3)$$

where $P_{av}(B,f_c,T)$ is the average power in the band of frequencies centered at f_c of width B that has been averaged for a time T. From (4-3) it should be apparent that one cannot just determine and record power without also knowing and recording its bandwidth. That is why, in general, one must plot the frequency spectra of a process as a power density (or its square root) so that the resulting amplitudes are normalized to account for the analysis bandwidth.

The choice of T and B is not independent for a given or desired statistical error on $G(f_c)$. If ϵ is the percentage normalized standard error of the power in band B averaged for time T, we find that they are related as

$$\epsilon = \frac{100}{\sqrt{BT}} \qquad \% \qquad\qquad (4\text{-}4)$$

A typical plot of power spectral density for a random process (signal) would look like that of Figure 4-2 except that $G(f)$ replaces $S(f)$. If one were measuring sound pressure in units of μPa, the ordinate of the power spectral density plot would be proportional to $(\mu\text{Pa})^2/\text{Hz}$.

4.2   FILTERS

Introduction

A filter is an electrical network having a pair of input terminals and a pair of output terminals. It may contain resistors, capacitors, inductors, or other components. When a time-varying voltage or "input signal," $s_i(t)$, is applied across the input terminals, a voltage, $s_o(t)$, known as the "output signal," is measured across the output terminals. If the input signal is a sine wave of frequency of f and amplitude A_i it can be shown that, the output amplitude A_o is different from A_i and that the sine wave has been delayed going through the filter, that is, there is a phase difference between $s_i(t)$ and $s_o(t)$. If we take the ratio of $s_o(t)$ to $s_i(t)$ at all possible values of frequency, we obtain the amplitude frequency response function of the filter, denoted $|H(f)|$, and the phase frequency response function denoted $\theta(f)$. It can be shown that $|H(f)|^2$ is proportional to the amount of energy (or power) that can be transmitted through the filter. Three basic types of filters are: low-pass, bandpass, and high-pass. These are shown in Figure 4-3.

To describe a filter, several of its characteristics must be specified. Consider the frequency response of a bandpass filter shown in Figure 4-4.

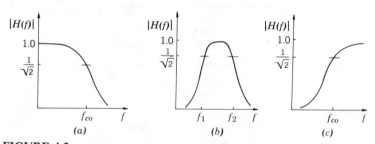

FIGURE 4-3
Three types of filters: (a) low-pass, (b) bandpass, and (c) high-pass.

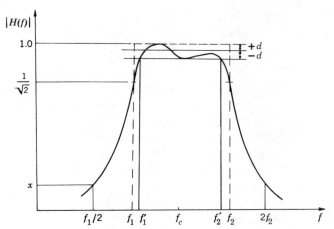

FIGURE 4-4
Frequency response of a bandpass filter. Ideal filter: (----).

The first characteristic is the *cutoff* frequencies of the filter. These are the points at which the $|H(f)|$ curve has decreased to $1/\sqrt{2} = 0.707$ of its peak value. The frequencies at which this occurs are called the cutoff frequencies of the filter. From Section 1.5 it is seen that this decrease in amplitude corresponds to 3 dB. The cutoff frequencies are also called the half-power points of the filter. In Figure 4-4 these are shown as f_1 and f_2, the lower and upper cutoff frequencies of the filter, respectively.

Another important characteristic of the filter is the rate of attenuation of the amplitude on either side of the cutoff frequencies. This descriptor, sometimes called the filter "skirts," is described by the number of dB/octave. From Figure 4-4 it would be obtained from $20 \log_{10}(0.707/x)$. The variation of $|H(f)|$ within the filter band, sometimes called filter ripple, would be specified as $\pm d$ from f_1' to f_2'. This specification is only meaningful for relatively broadband filters (typically, 1/3-octave and greater). A fourth description is the Q (quality factor) of the filter given by

$$Q = \frac{f_c}{f_2 - f_1} = \frac{f_c}{B} \tag{4-5}$$

As can be seen, it is possible for two seemingly similar filters to transmit different amounts of power. To differentiate still further, a term called the effective or noise bandwidth, B_e, of a filter is introduced. It is defined as the product of the 3 dB bandwidth and the ratio of the area described by $|H(f)|^2$ to that of an ideal filter. An ideal filter has a selectivity (skirts) of $-\infty$ dB/octave. A general rule is that the greater the attenuation of the filter skirts, the smaller the error compared to an ideal filter. In other words, the effective bandwidth approaches the -3 dB bandwidth.

Practical Filters

Two of the simplest types of practical filters are the low-pass and high-pass RC filters. The R denotes the resistance and C the capacitance. These two filters are shown in Figure 4-5. The low-pass filter has a frequency response similar to that shown in Figure 4-3a and the high-pass to that shown in Figure 4-3c. The cutoff frequency f_{co}, of both of these filters is $f_{co} = 1/(2\pi RC)$. It is a relatively easy matter to show that when a low-pass RC filter is used at frequencies much greater than f_{co} it acts as an integrator. Conversely, for a high-pass RC filter at frequencies much less than f_{co}, the filter acts as a differentiator.

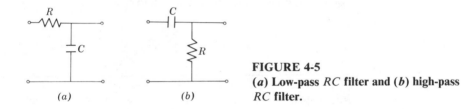

FIGURE 4-5
(*a*) **Low-pass** RC **filter and** (*b*) **high-pass** RC **filter.**

In general, filters can be classified as either a constant percentage bandpass filter, a constant bandwidth filter, or a variable bandwidth filter. These different filters are classified by the relationship of their center frequency to their cutoff frequencies. The center frequency is the geometric center between the cutoff frequencies. This is shown in (4-6). These filters are further classified depending on how they are actually constructed. If a frequency range is divided into N frequency bands (not necessarily equal) and if a filter is constructed so that it consists of N filters, each of which is responsive to frequencies only in that band, the filter is termed a *contiguous* filter. If the filter, however, only consisted of one filter whose center frequency is continuously variable through the frequency range, the filter is termed a *continuous* filter.

An example of a constant percentage bandpass filter of the contiguous type is the *octave* filter. It is a filter with the following characteristics [recall (1-43)–(1-45)]:

$$f_1 = 2^{-n/2}f_c \qquad f_c = \sqrt{f_1 f_2}$$
$$f_2 = 2^{n/2}f_c \qquad f_2 = 2^n f_1 \tag{4-6}$$

where f_1 and f_2 are the lower and upper cutoff frequencies, respectively, f_c is the center frequency, and n is the number of octaves. If $n=1$ we have a 1/1-octave filter; if $n=1/3$ we have a 1/3-octave filter. The bandwidth of the filter, B, is [recall (1-46)]

$$B = f_2 - f_1 = (2^{n/2} - 2^{-n/2})f_c = \beta f_c \qquad (4\text{-}7)$$

where β is the fraction (percentage) of the center frequency. Since n is a constant for a given type of filter, β is a constant, hence the classification constant percentage bandwidth filter. If $n=1$, $\beta=0.707$ or 71%; if $n=1/3$, $\beta=0.231$ or 23%.

Since 1/3- and 1/1-octave filters are the most commonly used octave filters, let us examine some of their properties. Consider Figure 4-6 which shows three contiguous 1/3-octave filters. Note that three contiguous 1/3-octave filters equal one 1/1-octave filter. Thus if P^I is the average power measured at the output of filter I, P^{II} the average power from filter II, and P^{III} the average power from filter III, then the average power in the 1/1-octave band is $P_{oct} = P^I + P^{II} + P^{III}$. Second, it is seen that the bandwidth of each succeeding 1/3-octave filter is, from (4-7), $2^{1/3}$ times as large as the preceding one, since each succeeding center frequency f_c is $2^{1/3}$ times as large as the preceding one. Thus the frequency resolution of each succeeding contiguous 1/3-octave filter is decreasing by $2^{1/3}$.

The octave filters are not used to perform narrow-band frequency analysis because their bandwidths are too wide, especially at higher ($f_c > 1000$ Hz) frequencies. Instead, narrow-band constant percentage continuous filters or narrow-band constant bandwidth filters are used. From

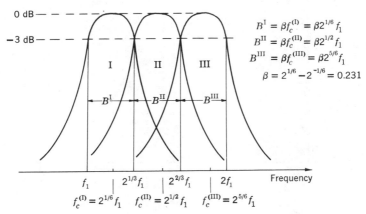

FIGURE 4-6
Three contiguous 1/3-octave filters.

(4-7) it is found that for a given β one can determine the equivalent number of octaves, n, from the following relation:

$$n = 6.64 \log_{10}\left(\frac{\beta}{2} + 0.5\sqrt{\beta^2 + 4} \right) \qquad (4\text{-}8)$$

For the case of a constant bandwidth analyzer, replace β by B/f_c. A constant bandwidth analyzer could either be a contiguous or continuous type. Most frequently they are of the latter type, although some of the so-called "real-time" analyzers employing digital techniques use them in a contiguous fashion.

As noted in the discussion following (4-3) it is usually necessary to divide the average power from each filter by its bandwidth. However, the center frequencies of each set of n-octave filters has been standardized by international agreement. (Recall Table 1-3). Consequently the bandwidths of each set of filters is the same and the division is not necessary. The only errors introduced between filter sets are due to the selectivity of the filters and the averaging time [recall (4-4)] used to obtain the level in each band. Regarding filter selectivity, it should be pointed out that there are three classes of octave, 1/2-octave, and 1/3-octave filters available as stipulated by a U. S. standards document.[2] Figure 4-7 shows their relative minimum and maximum allowable responses for the 1/3-octave filter. In addition, the peak-to-valley ripple of a Class II 1/3-octave filter may not exceed 1 dB. For a Class III filter this tolerance is reduced to 0.5 dB.

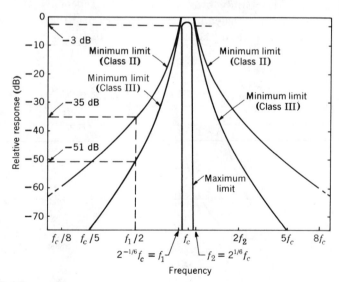

FIGURE 4-7
Amplitude response curves for Class II and Class III 1/3-octave filters. (From Reference 2.)

To illustrate the importance of filter selectivity, consider the following case: a signal having a single-frequency tone that is 10 dB higher than the levels in the adjacent 1/3-octave bands. This is not uncommon with machinery noise, for example. Assume that the frequency of the tone coincides with the center frequency of the 1/3-octave band. Now consider an adjacent 1/3-octave band. From Figure 4-7 it is seen that the Class II filter will attenuate the amplitude of this tone by at least 12 dB, whereas a Class III filter will attenuate it by at least 16 dB. Since we have assumed that the level of the tone is 10 dB higher than the noise level in the adjacent bands, the net attenuation is 2 dB and 6 dB, respectively. Consequently, for this example the level reading in the adjacent band will be approximately 2.1 dB higher for a Class II filter and 1.0 dB higher for a Class III filter. These values were obtained using Figure 1-9. Thus it is good practice to use filters with the highest selectivity possible to minimize the errors due to "amplitude leakage" from adjacent bands.

Before leaving the section on filters it is worthwhile to examine the errors associated with the use of practical filters to obtain estimates of the power spectrum. We ask the question, "What statistically significant results can be obtained in a minimum period of time with a given error?" If R_s is the scan rate, it can be shown that for RC averaging (integration)

$$R_s \leqslant \frac{B}{4T} = \frac{B^2\epsilon^2}{4} \tag{4-9}$$

wherein we have used (4-4). (Note that ϵ in this formula is expressed as a decimal number, not as a percentage). It can be shown that if (4-9) is not observed, two errors occur: a decrease in the peak amplitude of the output of the filter and an increase or widening of B. This will not occur if $R_s/B^2 \ll 1$.

The total analysis time, T_s, required to analyze a given signal over a frequency range $f_r = f_H - f_L$ for a constant percentage bandwidth filter is

$$T_s \geqslant \frac{4f_r BT}{f_H f_L \beta^2} = \frac{4f_r}{f_H f_L \epsilon^2 \beta^2} \quad \text{sec} \tag{4-10}$$

and for a constant bandwidth filter

$$T_s \geqslant \frac{4Tf_r}{B} = \frac{4f_r}{B^2\epsilon^2} \quad \text{sec} \tag{4-11}$$

Taking the ratio of (4-10) and (4-11) it can easily be seen that the total analysis time using constant percentage bandwidth filters will always be less than the total analysis time using constant bandwidth filters provided that $B < \beta\sqrt{f_H f_L}$.

Example 4-1.   For a 1/3-octave filter centered at 2000 Hz determine its (a) cutoff frequencies, (b) bandwidth, and (c) Q.

Solution: From (4-6) the upper and lower cutoff frequencies are

$$f_2 = (2^{1/6})(2000) = 2245 \text{ Hz}$$

$$f_1 = (2^{-1/6})(2000) = 1782 \text{ Hz}$$

The bandwidth is $B = 2245 - 1782 = 463$ Hz. This can also be obtained from (4-7); that is, $B = (0.231)(2000) = 462$ Hz, which agrees with the results presented in Table 1-3. The Q of the filter is obtained by combining (4-5) and (4-7) to yield

$$Q = \frac{1}{\beta} = \frac{1}{0.231} = 4.33$$

Example 4-2.   For the filter in Example 4-1 determine the minimum averaging (integration) time so that the normalized standard error is less than 10%.

Solution: Using (4-4) and the results of Example 4.1 yields

$$T \geqslant \left(\frac{100}{\epsilon}\right)^2 \frac{1}{B} = \left(\frac{100}{10}\right)^2 \frac{1}{463} = 0.216 \text{ sec}$$

Example 4-3.   What is the minimum total analysis time required to perform a frequency analysis using a continuous 6% bandwidth filter from 50 to 11,000 Hz while maintaining a statistical accuracy of 10%.

Solution: Using (4-10) yields

$$T_s \geqslant \frac{(4)(10,950)}{(50)(11,000)(0.1)^2(0.06)^2} = 2212 \text{ sec} = 36.9 \text{ min}$$

Example 4-4.   What is the minimum total time required to perform a frequency analysis over the range given in Example 4-3 using constant bandwidth filters having the same frequency resolution at 500 Hz and maintaining the same statistical accuracy.

Solution: The bandwidth of the 6% filter of Example 4-3 at 500 Hz is 30 Hz. Hence the bandwidth of the constant bandwidth filter is 30 Hz. Using (4-11) yields

$$T_s \geqslant \frac{(4)(10,950)}{(30)^2(0.1)^2} = 4867 \text{ sec} = 81.1 \text{ min}$$

Thus the 6% analysis is more than twice as fast. However, the frequency resolution for the constant bandwidth analysis is better above 500 Hz.

4.3 AMPLIFIERS

Introduction

Amplifiers have two basic functions. The first is to amplify signals that are too low in level for their intended application, and the second is to isolate circuits from other circuits. In performing these general functions the amplifiers introduce noise, that is, any spurious or undesired disturbance that tends to obscure or mask the input signal to the amplifier. A ratio found useful in describing the amount of noise contributed by an amplifier is the noise figure, NF, given by

$$NF = \frac{(S/N)_i}{(S/N)_o} \qquad (4\text{-}12)$$

where $(S/N)_i$ is the input signal-to-noise ratio and $(S/N)_o$ is the output (from the amplifier) signal-to-noise ratio. These ratios are on a power basis and the noise power (N) is without the signal. Since the amplifier will always add noise to the input signal, $(S/N)_o \leqslant (S/N)_i$ and, therefore, $NF \geqslant 1$. When (4-12) is expressed in decibels we have $10 \log_{10}(NF)$. The noise figure is a function of frequency and the resistance of the source.

Many times it is necessary to use more than one amplifier to increase the level of the input signal. When one amplifier is followed by another amplifier they are said to be cascaded. If the first amplifier has a gain A_1, and a noise figure NF_1, and the second amplifier has a gain A_2 and a noise figure NF_2, it can be shown that the total noise figure for the cascaded amplifiers, NF_T, is

$$NF_T = NF_1 + \frac{NF_2 - 1}{A_1} \qquad (4\text{-}13)$$

It is clear that the relative contribution of the two cascaded amplifiers is most dependent on the last amplifier. Hence for the best overall noise figure the first amplifier should have the lowest possible noise figure and the highest possible gain.

Another indicator of the capabilities of an amplifier is its dynamic range. The dynamic range is the region between the inherent noise of the amplifier and the amplitude level beyond which a specified percentage harmonic distortion occurs. Thus if a manufacturer states that the amplifier noise is 10 mV (referred to input), its dynamic range 60 dB, and its percentage harmonic distortion 1%, the maximum permissible input voltage would be 10 V. (The term "noise referred to input" is the noise measured at the output of the amplifier divided by the gain of the amplifier.) In this example, the amplifier could be used with voltage between 10 mV and 10 V. To increase the range of voltages an attenuator usually precedes the actual amplifier. The purpose of the attenuator is to reduce the magnitude of the input signal so that it falls within the dynamic range of the amplifier. It should be noted that the attenuator has not increased the dynamic range of the amplifier, but has increased only the range of voltages with which the amplifier may be used.

Impedance Matching

The introduction of any measuring instrument into the measured medium always results in the extraction of some energy from that medium. This extraction of energy changes the value of the measured quantity from its undisturbed state and introduces an error in the measurement. To minimize this error almost no current should flow from the measured device to the measuring device. This is what is meant by using an amplifier to isolate one circuit from another. The rule is that the output impedance (resistance) of one device (transducer, amplifier, etc.) should always be at least 100 times less than the input impedance of the device to which it is connected (the only exception is given at the end of this section). Consider the following two cases shown in Figure 4-8: (a) a resistive load R_g connected to the input of an amplifier having an input resistance R_i and a shunt capacitance C_i; (b) a capacitative load C_g connected to the input of an amplifier having an input resistance R_i and shunt capacitance C_i. For the former case the percentage error is given by

$$\% \text{ error} = 100\left\{1 - \left[(1+\alpha)^2 + \alpha^2\omega^2\tau_i^2\right]^{-1/2}\right\} \% \qquad (4\text{-}14)$$

where $\alpha = R_g/R_i$, $\tau_i = R_i C_i$ and ω the circular frequency. For the latter case we find that the percentage error is given by

$$\% \text{ error} = 100\left\{1 - \beta\omega\tau_i\left[1 + \omega^2\tau_i^2(1+\beta)^2\right]^{-1/2}\right\} \% \qquad (4\text{-}15)$$

where $\beta = C_g/C_i$. Comparing Figure 4-5 with Figure 4-8 shows that the configuration in Figure 4-8a forms a low-pass filter and that shown in Figure 4-8b forms a high-pass filter. Equations (4-14) and (4-15) are plotted in Figures 4-9 and 4-10, respectively.

Figure 4-9 illustrates that in order to have a very high cutoff frequency with the smallest error $\alpha = R_g/R_i \ll 1$ and $\alpha\tau_i = R_g C_i \ll 1$. This last inequal-

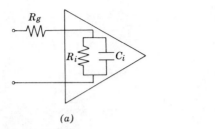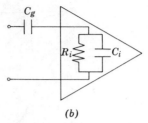

(a)   (b)

FIGURE 4-8
(a) **Resistive load and** (b) **capacitive load connected to the input of an amplifier having an input resistance R_i and a shunt capacitance C_i.**

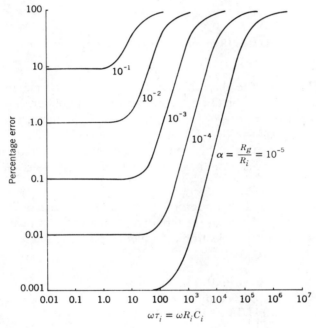

FIGURE 4-9
Plot of (4-14) for various α. (Reprinted, by permission, from Reference 1.)

ity requires that C_i be as small as possible. Figure 4-10 shows that for minimum error at low frequencies C_g/C_i should be as large as possible. In addition, in order to have a very low cutoff frequency with minimum error, $R_i C_i \gg 1$. However, it can be shown that if C_i is so large that C_g/C_i decreases too much, the amplitude of the input signal is also attenuated. Thus R_i must have a large value, so that the above inequality is satisfied.

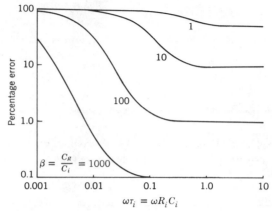

FIGURE 4-10
Plot of (4-15) for various β. (Reprinted, by permission, from Reference 1.)

The one exception to the rule that $R_g/R_i \ll 1$ is when maximum power transfer is required to a load on the output of the amplifier. In this case, if R_0 is the output resistance of the amplifier and R_L is the resistance of the load, it can be shown that for maximum power transfer, $R_L = R_o$. In addition, if the load is highly reactive, the reactive part of the output impedance must equal the negative of the reactive part of the load impedance.

Example 4-5.   A condenser microphone has a capacitance of 50×10^{-12} F. What should be the values of the input resistance and shunt capacitance of the amplifier so that the electronic frequency response has less than 0.1% error from 20 to 20,000 Hz.

Solution: From Figure 4-10 it is seen that when $C_g/C_i = 1000$ and $\omega R_i C_i = 0.1$ the percentage error will be 0.1%. Thus $C_i = 50 \times 10^{-12}/1000 = 0.05 \times 10^{-12}$ F and $R_i = (0.1)/[(2\pi)(20)(0.05 \times 10^{-12})] = 16 \times 10^9$ Ω.

Transient Response of Bandlimited Systems

Consider a linear system (amplifier, filter, transducer, etc.) that has cutoff frequencies f_1 and f_2 ($f_2 > f_1 > 0$) so that the bandwidth is $B_r = f_2 - f_1$. We define the rise time, τ_r, as the time it takes the system, in response to a sudden change in amplitude, to go from 10 to 90% of the final or steady-state value. It can be shown that the rise time of a bandpass system is

$$\tau_r \approx \frac{1}{B_r} \tag{4-16}$$

To illustrate the significance of this result, consider a typical response of the envelope (the actual waveform does not matter) of a linear system, as shown in Figure 4-11. If this system is subjected to a signal whose duration is τ_1, three possible cases can be examined. The first is $\tau_1 \ll \tau_r$ $(B_1 \gg B_r)$. In this case the system has very little time to build up, hence the system's output will be very small and resemble the input waveform very little. The second case is $\tau_1 \approx \tau_r (B_1 \approx B_r)$. Here the output amplitude will be very close to the steady-state value. Its waveform will be somewhat definable but it will only crudely resemble the actual input signal. The third case is when $\tau_1 \gg \tau_r$ $(B_1 \ll B_r)$. In this case the system has sufficient time to respond and the output waveform is almost an exact replica of the input waveform. It can be shown that the high frequency cutoff should be at least greater than or equal to $10/\tau_0$, where τ_0 is the approximate duration of the steepest ascent of the pulse (transient) waveform. Although not as obvious from the presentation herein, there is also a low frequency requirement that $f_1 \leqslant 0.008/\tau_0$. If this low frequency is not adhered to there will be an undershoot at the end of the pulse.

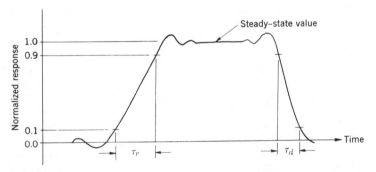

FIGURE 4-11
Typical response of the positive-going envelope of a band-limited linear system.

Finally, when two or more bandlimited linear systems are cascaded, larger bandwidths for each individual system are required to obtain an overall rise time equal to that of a single system. As a rule of thumb two cascaded systems each require a bandwidth of 1.5 times the bandwidth of one system, whereas three cascaded systems require a bandwidth of 1.9 times the bandwidth of one system. For example, if a rise time of 1 msec is desired for the output of two cascaded amplifiers, each amplifier should have a bandwidth of approximately 1500 Hz.

4.4 VOLTAGE DETECTORS

The majority of signals in measurement systems ultimately appear as voltages. Since voltage cannot be seen, it must be converted into a form intelligible to the observer. One way to accomplish this is to convert the time-varying voltage into a slowly varying dc voltage that is proportional to the input voltage. The dc voltage is then displayed by a meter or recorder. The converter in this case is called the voltage detector. There are three types of detection: peak, average, and rms.

A peak detector gives the maximum value (either positive or negative) during some time interval T. An average detector measures the mean absolute (or rectified) value of the signal during a time T. A rms detector measures the square root of the mean value of the square of the signal during a time interval T. [See (4-1)]. A term used to describe the capability of a rms detector is the *crest factor* defined as the ratio of the peak value of the signal to its rms value. It can be thought of as the measure of the upper limit of the dynamic range of the detector. The larger this value, the greater the dynamic range.

TABLE 4-1

Comparison of Rectified Average and Peak Detection to the RMS Value for Various Types of Signals[a]

Waveform	RMS meter indicates	Rectified average meter indicates (Calibrated in rms)	Peak meter indicates (Calibrated in rms)	Error (dB) $[20\log_{10}(E/E_{rms})]$ Average	Peak
Sine wave	0.707	0.707	0.707	0	0
Sine wave Plus 100% Third harmonic					
In phase	1.000	0.944	1.09	−0.50	0.748
Out of phase	1.000	0.472	0.382	−6.52	−8.36
Square wave	1.000	1.111	0.707	+0.91	−3.00
Gaussian noise	σ_0	$0.887\sigma_0$	—	−1.04	—
Pulse train					
$D=0.1$	0.318	0.111	0.707	−9.14	+6.10
$D=0.01$	0.10	0.011	0.707	−19.17	+16.99

[a]Reprinted, by permission, from Reference 1.

It is common practice for many commercial detectors to use scales that are calibrated to read the rms voltage of a *sine wave*. However, if the detector is not a rms one, an error will occur. Table 4-1 depicts what the peak, rectified average, and rms meter will read when calibrated in rms for a sine wave, and what the error in dB is compared to the true rms reading. In the table, σ_0 is the standard deviation of the Gaussian process and D is the ratio of the pulse duration to the period. The message from Table 4-1 is clear; use only a true rms detector.

4.5   MICROPHONES[3,4]

Introduction

There are numerous factors, such as frequency response, directivity, dynamic range, stability, sensitivity, temperature, humidity, and wind, that must be considered before one can make the proper selection of a microphone. These aspects are discussed in detail in this section.

Condenser Microphone Construction

The schematic view of a typical condenser microphone is shown in Figure 4-12. The microphone consists of a thin metallic diaphragm mounted in close proximity to a rigid backplate. The diaphragm and backplate are electrically insulated from each other and constitute the electrodes of a capacitor. The housing and insulator form, with the diaphragm, a closed chamber. To compensate for slow static ambient pressure variations a small capillary tube connects the chamber with the outside. When the microphone is exposed to a sound pressure, the diaphragm is submitted to an alternating force proportional to the pressure and the diaphragm area. The resulting movement of the diaphragm varies the capacitance between the backplate and the diaphragm. These variations are transduced into an alternating voltage since a constant charge is present between the electrodes. The charge is obtained by means of a stabilized dc polarization voltage.

The widest linear frequency range for the pressure response is obtained if the resonance of the mechanical system (diaphragm) is critically damped. This damping, which is caused by the back and forth motion of the air contained between the diaphragm and backplate, is determined by the shape of the backplate, the mechanical tension of the diaphragm, and the distance between the diaphragm and the backplate. The low frequency limit of the condenser microphone is determined by the electronics immediately following the microphone cartridge. [Recall (4-15).]

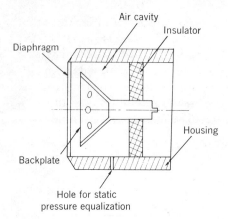

FIGURE 4-12

Schematic view of a typical condenser microphone. (From Reference 3.)

The frequency range of the condenser microphone can be extended by decreasing its size (diameter of the diaphragm). For example, if the response of a "1-inch" microphone is linear (within ±1 dB) to 18 kHz, a "1/2-inch" microphone will be linear to approximately 40 kHz, and a "1/4-inch" microphone to over 100 kHz. However, as size is decreased, so is sensitivity. For example, going from a "1-" to a "1/2-inch" microphone, the sensitivity decreases 14 dB; from a "1/2-" to a "1/4-inch", 20 dB.

Other useful features of condenser microphones are their long-term stability and almost uniform sensitivity as a function of temperature. If the microphone has undergone accelerated aging, the condenser microphone will typically change its sensitivity less than 0.1 dB every 10 yr when operating below 35°C. The change in sensitivity as a function of temperature is typically 0.008 dB/°C between −50 and 150°C. The change with respect to humidity is usually less than 0.5 dB between 0 and 90 % relative humidity. The effect of static (ambient) pressure is −0.003 dB/mm Hg.

One drawback of a condenser microphone is its sensitivity to moisture. When a microphone is moved from a warm environment to a cold one, moisture condensation may take place. In this case it is important that the pressure equalization arrangement connect the inside of the microphone directly to the outside atmosphere. A microphone cartridge will always contain a certain amount of water vapor, depending on the temperature and the relative humidity of the air. If the microphone is cooled down the relative humidity will increase very rapidly and soon reach 100%, where condensation will occur. This condition may result in a noisy microphone. This condensation of moisture will take place in those parts of the microphone that are cooled down first, which are primarily the free parts of the diaphragm between the backplate and the rim. In the beginning, at least, no condensation will take place between the diaphragm and the backplate, for the backplate is the last part of the microphone to be cooled down.

On the other hand, when a cool microphone is brought into a hot environment some of the moisture in the hot air will diffuse into the microphone via the equalization port causing moisture condensation on the coldest places, which are between the backplate and the diaphragm. This almost invariably results in noise. These condensation problems have been eliminated in certain condenser microphone designs, however, by changing the location of the pressure equalization port to go through the insulator to the back of the microphone. Instead of having the amplifier immediately following the microphone, a small chamber is inserted between the microphone and its electronics. This chamber contains silica gel, which effectively dries the air used by the microphone for pressure equalization. The other end of this chamber contains the pressure equalization port to the atmosphere.

Electret Microphone Construction

The schematic view of a typical foil electret microphone is shown in Figure 4-13. The nonmetallized surface of the foil electret is placed next to a backplate, leaving a shallow air gap whose thickness (about 10 μm) is controlled by ridges or raised points on the backplate surface (not shown in the figure). The backplate is either a metal disk or a metal-coated dielectric having a thermal expansion coefficient about equal to that of the foil. To decrease the stiffness of the air layer and thus improve the microphone sensitivity, it is connected to a larger air cavity by means of small holes through the backplate. The electrical output of the microphone is taken between the backplate, which is insulated from the outer case, and the metal side of the foil. Thus the electret microphone is a condenser microphone with a solid dielectric for its diaphragm; it differs from the condenser microphone by the fact that it does not require a dc polarization voltage. In addition the electret microphones contain most of the positive features of the condenser microphone. They differ considerably from the condenser microphones in their long-term stability and their temperature coefficient. Typical temperature coefficient for the electret microphone is 0.03 dB/°C from 0 to 55°C. Regarding long-term stability some results[4] show that over a 15-month period during which the electrets were stored at room temperature the sensitivity varied between ±1 dB. Other results[5] show even greater variations. These temperature and stability properties of electret microphones may not permit consistent and repeatable measurement results. Furthermore, if the electret is used to perform community noise measurements (L_{eq}, L_{dn}, etc.) errors may be introduced if the environment in which the measurement is being made undergoes moderate to large temperature variations (15–20°C).

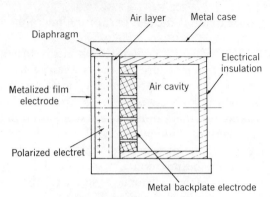

FIGURE 4-13
Schematic view of a typical foil electret microphone. (From Reference 4.)

Some electret microphones also differ from condenser microphones in their frequency range. A "1-inch" electret microphone typically has a uniform (within ± 1 dB) frequency response to approximately 12 kHz. The "1/2-inch" microphone increases this value to 15–20 kHz, whereas the "1/4-inch" microphone goes to approximately 30 kHz. These relatively narrow band-widths for the smaller microphones may be a limitation in measuring the peak levels of short duration events, such as those emitted by metal-to-metal impacts and small firearms. However, the sensitivity of the microphones does not decrease as drastically as that of a condenser microphone as the size is decreased. For example, going from a "1-" to a "1/2-inch" electret the sensitivity decreases 6 dB; from a "1/2-" to a "1/4-inch", it decreases 15 dB.

Sensitivity and Dynamic Range

The sensitivity of a microphone is the minimum detectable change in output voltage for a change in the acoustic pressure impinging upon the microphone. It is usually expressed in dB re $1V/\mu Pa$. The less negative the value of this level, the greater is its sensitivity.

The least sound-pressure level that a microphone can respond to is governed by the internal noise of the electronics that follow and can be as low as an A-weighted level of 11 dB. The upper limit is sometimes defined as the level at which 3% harmonic distortion occurs. This is the total distortion due to both the microphone diaphragm and the electronics. For "1-inch" condenser microphones this value can be as high as 145 dB, for "1/2-inch" microphones 160 dB, and for "1/4-inch" microphones 175 dB. Typically, if these 3% distortion levels are exceeded by sound pressures that are 10–15 dB higher, the condenser microphone diaphragm will be damaged. For "1-inch" electret microphones the upper limit is typically

140 dB, for "1/2-inch" microphones 145 dB, and for "1/4-inch" microphones 150 dB. If these levels are exceeded by approximately 20 dB the electret may be damaged.

Frequency Response

If a microphone is placed in a sound field consisting of a plane wave that has a pressure p_0, the sound wave will be partly reflected from the microphone, causing an increase in the sound pressure at the microphone diaphragm. The magnitude of this increase depends on the wavelength of the sound, the physical dimensions of the microphone diaphragm, and the direction of travel of the sound with respect to the diaphragm. Often the objective is to measure the sound pressure that exists in a sound field before the microphone was placed in it. Hence we define the *free-field* response as the ratio of the output voltage from the microphone to the sound pressure of the undisturbed field, that is, at the microphone location with the microphone removed. The voltage and pressure in this ratio are rms values. We define the *pressure* response of a microphone as the ratio of the output voltage from the microphone to the sound pressure uniformly applied over the microphone diaphragm. Then the free-field response is equal to the pressure response plus the pressure increase caused by the reflections. When the pressure response curve is known, the problem of finding the free-field response curve is reduced to that of finding the pressure increase correction curve, which is dependent on the shape and physical dimensions of the diaphragm, the frequency of the sound, and the sensitivity distribution on the diaphragm. It is therefore necessary for a given microphone to determine the pressure increase correction curve, which is then added to the individual measured pressure response curve of the microphone. This correction becomes important when the ratio of the microphone diameter to the wavelength of sound becomes greater than 0.1 (moderately high frequencies). When the wavelength of the sound is very small compared to the diameter of the microphone diaphragm, the sound waves that are perpendicularly incident to the diaphragm will be reflected as if they were impinging on a wall of infinite dimensions. A pressure increase of 6 dB is obtained. When the impinging sound wave is parallel to the plane of the diaphragm, the output voltage of the microphone will approach zero at high frequencies.

In the range between very low frequencies (where the pressure increase is negligible) and very high frequencies (where the pressure increase is 6 dB) a varying frequency response is obtained, depending on the different resonances at the microphone surface. The pressure increase given in dB relative to the sound pressure existing before the microphone was placed in the sound field is the basis for a family of experimentally determined curves, known as the *free-field corrections*, which should accompany each microphone. They are the same for all microphones having the same diameter, diaphragm properties, and shape. As stated before, these corrections are added to the pressure characteristics of the microphone. The free-field correction curves are a function of the angle of incidence of the impinging plane waves. Figure 4-14 shows typical free-field correction curves for a "1-inch" condenser microphone. Figure 4-14 can be replotted to yield typical directivity characteristics of these microphones, as shown in Figure 4-15.

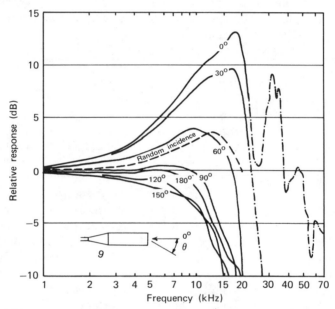

FIGURE 4-14
Typical free-field correction curves to be added to the pressure response curve of a "1-inch" microphone. (From Reference 3.)

Figure 4-14 shows a correction curve marked "random incidence." The random incidence response of a microphone for a given frequency is the rms value of the free-field sensitivity for all angles of incidence of the sound wave. It corresponds to the diffuse field sensitivity of the microphone. As noted previously, the diffuse field is a sound field in which the sound energy density is uniform and the mean acoustic power per unit area is the same in all directions. This curve is usually determined from an approximate formula.[6]

Rather than performing this elaborate correction, if it is known that the

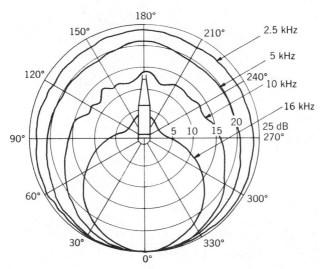

FIGURE 4-15
Typical directional characteristics of the microphone of Figure 4-14. (From Reference 3.)

sound incident upon the microphone comes from all directions (a diffuse sound field), the random incidence correction should be as close to zero as possible over the widest frequency range. This can usually be accomplished with either a random incidence corrector or a nose cone. The resulting free-field corrections for a random incidence corrector are shown in Figure 4-16. Compare this figure with Figure 4-14. Another way to interpret Figure 4-16 is to realize that both of these attachments give the microphone a more omnidirectional character; that is, the curves in Figure 4-15 would almost be circles at all frequencies up to 10 kHz with a maximum deviation from a circle of 5 dB.

Another use for the nose cone is to reduce noise produced by turbulence when the microphone is exposed to high wind speeds; it is especially effective with constant wind speeds of fixed direction.

For all outdoor noise measurements it is advisable to use a *windscreen*. The material of the windscreen presents a great resistance to the wind but the acoustic impedance remains small. Unfortunately, most manufacturers' supplied windscreens are virtually useless in winds above 10 km/hr. Recently work[7] has been performed on windscreens that permit microphones to be used in winds up to 40 km/hr. If an 18-cm diameter sphere of open-cell polyurethene having a linear pore density of 800 pores/m is used, the overall A-weighted noise level due to the 40 km/hr

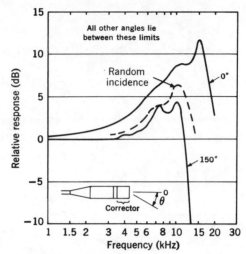

FIGURE 4-16
Typical free-field corrections of the microphone of Figure 4-14 with a random incidence corrector. (From Reference 3.)

wind is less than 55 dB and the 1/3-octave band pressure levels are less than 46 dB from 100 to 10000 Hz. At lower wind speeds these levels are correspondingly lower. The attenuation (insertion loss) caused by the sphere is negligible.

Microphone Orientation

Two types of condenser microphones are available for making measurements: the pressure microphone and the free-field microphone. They are constructed in exactly the same manner except that the shape of the backplate and its distance from the diaphragm are different in each. This gives different frequency response curves for each type as a function of angle of incidence. If these backplate parameters are carefully designed, it is possible to build in the free-field correction at a particular angle of incidence; that is, the sum of the pressure response curve of the microphone and its correction curve at a specified angle of incidence will give a uniform (linear) free-field response at that angle of incidence. The two angles selected are 0°, perpendicular incidence, and 90°, grazing incidence. The modified response of the microphone and its free-field corrections are shown in Figure 4-17. This procedure for building microphones removes the need to apply the free-field correction if the angle of incidence is known, since the microphone can then be oriented to give the most uniform frequency response over the widest frequency range.

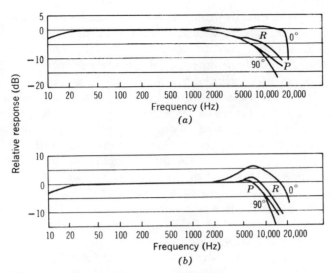

FIGURE 4-17
**Perpendicular (0°) and grazing (90°) incidence frequency response of (a) a free-field
condenser microphone and (b) a pressure condenser microphone. *R*—Random incidence response. *P*—Pressure response. (From Reference 3.)**

As their names imply, the pressure microphones are used predominantly
under close-coupled or constant pressure-measuring conditions, such as
those in the near-field of a sound source. The free-field microphone is most
often used under free-field or semireverberant conditions. There are five
common conditions under which measurements are performed. They are
perpendicular incidence, omnidirectional, pressure, grazing incidence, and
random incidence. These are shown in Figure 4-18.

Perpendicular (0° incidence) free-field measurements are made when

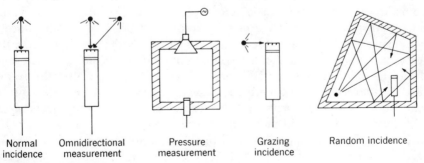

Normal   Omnidirectional   Pressure   Grazing   Random incidence
incidence   measurement   measurement   incidence

FIGURE 4-18
Five common measurement conditions for microphones.

the sound radiated directly from the source is to be determined. This is only possible if there are no reflecting surfaces in the vicinity of the sound source and microphone. An omnidirectional microphone is one that responds to sounds equally, irrespective of the angle of incidence. This type of microphone is used when the total sound from many directions is desired, for example, noise measured indoors, noise from several sources simultaneously, and street noise. To perform this type of measurement one selects either the smallest microphone that will still give the necessary sensitivity or uses the random incidence corrector. With the random incidence corrector a correction for the free-field will have to be made if there is a considerable amount of high-frequency sound. By using a small microphone the frequency response is more uniform over an extended frequency range, thus the free-field correction is not ordinarily required. It should be noted that these free-field corrections would have to be applied after a frequency analysis has been performed.

A pressure microphone provides the most uniform frequency response when closely coupled to the sound. This type of microphone is used to calibrate earphones and audiometers, regulate sound sources, and probe the near field of a sound source. A pressure microphone should also be used in its grazing incidence orientation where the sound from a moving source, such as an automobile, boat, or airplane on a runway, is being measured. By using the pressure microphone in its grazing orientation the response of the microphone is the same, irrespective of the position of the moving object.

A pressure microphone is again used in a reverberant room when the objective is to measure the total sound power from a noise source. The random incidence response of the pressure microphone is relatively uniform over a very broad frequency range.

Sound Level Meter

The sound level meter is a battery operated device that contains the appropriate electronics to convert the sound pressure exciting the microphone diaphragm into a meter reading, in dB re 20 μPa. The concern with using a portable sound level meter is the effect the person holding the meter has on the sound field. Certain combinations of noise spectra and distances of the sound level meter from one's body can alter the sound field by several dB. For precision measurements these effects can be eliminated by simply placing a 2–3 m cable between the sound level meter and the microphone and mounting the microphone on a tripod.

For many measurements it is sufficient to simply hand hold the sound level meter and record the levels directly from the meter reading. However, under certain circumstances it is necessary for the sound level meter to have additional capabilities. To record the noise using a portable tape recorder the sound level meter must have an ac output terminal. In addition, in order to calibrate the entire sound recording system a calibrator or pistonphone must be used. These aspects are discussed in detail on pp. 114 and 117.

For measuring vehicular pass-by levels or one-time, short duration (impulsive type) noise events a "hold" circuit is a necessity. The use of a hold feature is necessary for two reasons: (1) since the duration of the peak pass-by or impulse noise is often less than or equal to the response time of the meter, the hold circuit electronically holds the maximum value permitting the meter itself sufficient time to respond, and (2) since the peak sound-pressure level only lasts a brief moment, it would be difficult to accurately and consistently read the meter (if it could respond correctly) in such a short period of time.

Another added feature of the sound level meter is an input amplifier overload indicator light. A sound level meter will often be used to record the A-weighted-sound level. In many instances the difference between the A-weighted levels and the unweighted levels can be substantial (>7 dB). In these instances it is possible for the input amplifiers to be overloaded and, therefore, clip (limit the peak amplitudes of) the input signal even though the meter reading is not registering near its maximum value. This is even more likely to occur when one attaches an octave filter set to the sound level meter and performs an octave band analysis. Then there is an even greater difference between the octave band levels and the overall weighted or unweighted sound pressure level.If an overload indicator is not available or if it is inconvenient to monitor, the same function can be accomplished using a pair of high-impedance well-fitting earphones. Low impedance earphones will load the output of the sound level meter output circuit so that falsely low readings are obtained. (Recall Figure 4-8.)

To minimize the possibility of overloading the input amplifier the following procedure should be used with those sound level meters having a second or output amplifier stage. Set the weighting switch to "linear" or "C-weighting." Adjust the range attenuator until the meter reads "maximum" without going off scale. Switch in the A-weighting network or the filters and increase the meter deflection to as close to full scale as possible by adjusting the attenuator preceding the output amplifiers. Notice that the maximum "gain" (least attenuation) is used by the first amplifier, which is necessary to keep the overall noise figure at a minimum [recall (4-13)]. Overloading the first amplifier is a very common error in the usage of a sound level meter and the importance of the procedure described above cannot be overemphasized.

There are four classes[7] of sound level meters in common use: Type

1—precision, Type 2—general purpose, Type 3—survey, and Type S— special purpose. In certain instances sound level meters may be required to perform special purposes that do not require the complexity of any of the three basic types. Therefore, provision is made for a special purpose sound level meter, with the use of the designation Type S. The Type S meter can be qualified to the performance of any of the basic types (1, 2, and 3), but is not required to have all three weighting networks.

We now discuss some of the relative properties of Types 1–3. The tolerances for the A-weighting are shown in Table 4-2 and apply for sound at random incidence and include all tolerances in the entire sound level meter and microphone. The symbol $-\infty$ (minus infinity) in the table means that the instrument does not have to respond at all to this frequency; that is, its output may be zero.

To understand the significance of these tolerances consider the case of a noise having a constant power in each 1/3-octave band from 100 Hz to 4 kHz (pink noise). Then the range of error, that is, the difference in the errors using the maximum positive tolerances and the maximum negative tolerances given in Table 4-2, is as follows: For a Type 1 meter ($+1.2$ dB, -0.90 dB), for a Type 2 meter ($+3.4$ dB, -2.6 dB), and for a Type 3 meter ($+4.0$ dB, -3.7 dB). This example clearly illustrates the fact that two calibrated instruments of the same type can give readings that differ by a substantial amount, yet both meet the required specification tolerances.

Sound level meters possess two types of dynamic responses for their meter: fast and slow. These responses are defined as follows. If a tone with a frequency of 1000 Hz and a duration of 0.2 sec is applied, for fast response, the maximum reading for a Type 1 instrument is between 0 and 2 dB less than the reading for a steady signal of the same frequency and amplitude. For Types 2 and 3 the maximum reading is between 0 and 4 dB less. However, this requirement holds for steady readings 4 dB less than full scale. For slow response, the duration of the 1000 Hz tone is 0.5 sec and the maximum reading for a Type 1 instrument is 3–5 dB less than the reading for a steady signal of the same frequency and amplitude. For Types 2 and 3 the maximum reading is between 2 and 6 dB less. These figures should emphasize the points made in the previous discussion of the "hold" feature. The implication of the tone-burst response of a sound level meter is that it must be used with care when measuring noises that are impulsive in character, such as in factories using punch presses, drop-forges, or any type of machinery that has an intermittent cycle. The work by Wilkerson[8] clearly illustrates this point.

It is also pointed out that the above mentioned requirements characterize primarily the meter ballistics and do not necessarily indicate

TABLE 4-2

Total Tolerance Limits for A-Weighted Sound at Random Incidence for Sound Level Meter Types 1, 2, and 3[a]

Frequency (Hz)	Type 1	Type 2	Type 3
10	±4.0	—	—
12.5	±3.5	—	—
16	±3.0	—	—
20	±2.5	+5.0, −∞	+6.0, −∞
25	±2.0	+4.0, −4.5	+5.0, −6.0
31.5	±1.5	+3.5, −4.0	+4.5, −5.0
40	±1.5	+3.0, −3.5	+4.0, −4.5
50	±1.0	±3.0	±4.0
63	±1.0	±3.0	±4.0
80	±1.0	±3.0	±3.5
100	±1.0	±2.5	±3.5
125	±1.0	±2.5	±3.0
160	±1.0	±2.5	±3.0
200	±1.0	±2.5	±3.0
250	±1.0	±2.5	±3.0
315	±1.0	±2.0	±3.0
400	±1.0	±2.0	±3.0
500	±1.0	±2.0	±3.0
630	±1.0	±2.0	±3.0
800	±1.0	±1.5	±3.0
1,000	±1.0	±2.0	±3.0
1,250	±1.0	±2.0	±3.0
1,600	±1.0	±2.5	±3.5
2,000	±1.0	±3.0	±4.0
2,500	±1.0	+4.0, −3.5	±4.5
3,150	±1.0	+5.0, −4.0	±5.0
4,000	±1.0	+5.5, −4.5	±5.5
5,000	+1.5, −2	+6.0, −5.0	±6.5
6,300	+1.5, −2	+6.5, −5.5	±7.0
8,000	+1.5, −3	+6.5, −6.5	±7.5
10,000	+2.0, −4	+6.5, −∞	+7.5, −∞
12,500	+3.0, −6	—	—
16,000	+3.0, −∞	—	—
20,000	+3.0, −∞	—	—

[a]From Reference 6.

the type of detector (rms or rectified average) used. Consequently, one should not assume the type of detector a sound level meter has, but should instead look for a specific statement by the manufacturer. (Recall Section 4.4 and Table 4-1.)

Calibration of Microphones

The sensitivity of microphones is most practically determined by a mechanical device known as either a calibrator or a pistonphone. These devices generate a known sound-pressure level mechanically and can be as accurate as ± 0.2 dB. They generate a high level sound at a single frequency, usually between 90 and 125 dB, so that a microphone can be calibrated in relatively noisy environments. The frequency of the calibration tone is usually between 200 and 1250 Hz.

Effects of Background Noise

To determine whether the background noise level in a space influences the sound-pressure levels, one should turn off the noise source(s) and measure the background noise level. This measurement should be made with the same analysis system that was used when the noise source was on. If in any frequency band the difference between the background noise level and the source noise level is greater than 10 dB, the background level in that frequency band will not significantly affect measurement of the source noise. However, if the difference is less than 10 dB, the measured noise levels must be corrected to obtain the level of the source. As a practical matter, if this difference is less than 3 dB the measurement has very little chance of being reliable and therefore should not be recorded. The corrections to be applied are shown in Figure 4-19, which is simply a replotting of Figure 1-10. In Figure 4-19 ΔL_N is the correction to be subtracted from the sound-pressure level with the source on, denoted L_{S+N}, and L_N is the sound-pressure level of the background noise. For example, if the reading with the sound source on is 100 dB and 93 dB with it off, we see from Figure 4-19 that the correct reading for the source is 99 dB.

Calibration of a Tape Recorder and a Sound Level Meter

It is often necessary to record sounds in the field and then bring them back to the laboratory for further analysis. Thus one must be able to calibrate the entire system so that when the tape is played back, the voltage can be related to the acoustic reference level of 20 μPa. Consider the simplified schematic of a sound level meter and a tape recorder shown in Figure 4-20a. In Figure 4-20a, A_1 is a very accurate attenuator which

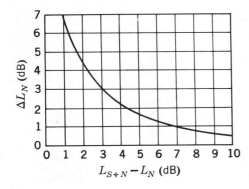

FIGURE 4-19
Correction chart for background noise.

decreases the signal from the microphone, usually in 10 dB increments, and A_2 is a continuously variable attenuator. The amplifiers are fixed-gain amplifiers, that is, when V_0 is an input, V_1 is always the output. If one were reading the meter and V_0 corresponded to the midpoint on the meter scale, for example, whenever the meter indicator was at this midpoint, V_1 would always be the output voltage of the sound level meter. The attenuator controlling the input voltage level to the record amplifier of the tape recorder is designated A_3.

To calibrate the system a calibrator or a pistonphone is placed over the microphone. This produces a′ known sound-pressure level of z dB re 20 μPa on the microphone, which in turn produces a voltage. The fixed input attenuator A_1 is adjusted until the meter, whose scale is also in dB, reads as close to full scale as possible. If the fixed attenuator setting indicates x dB and the meter y dB, the continuously variable attenuator A_2 is adjusted until $x + y = z$. Now A_2 must remain fixed in this position, otherwise calibration will be lost. With the pistonphone or calibrator still on the microphone, A_3 is adjusted by reading the meter on the tape recorder so that V_2 falls within the dynamic range of the system. As a maximum, V_2 should be w dB from the maximum allowable voltage, where w is the difference, in dB, between the sound level meter's full scale value and that produced by the pistonphone. However, since the type of detector used in the tape recorder may not be known and may differ from that on the sound level meter it is advisable to be conservative and let V_2 be $V + 2.0$ dB less than full scale. Under no circumstances should A_3 be changed during the entire recording process or calibration will be lost. The tape recorder is now started and about 10–20 sec of the pistonphone calibration tone is put on the beginning of the tape. The pistonphone is removed and the system is ready to record the sounds of interest.

When recording the sounds, the input attenuator A_1 is usually adjusted

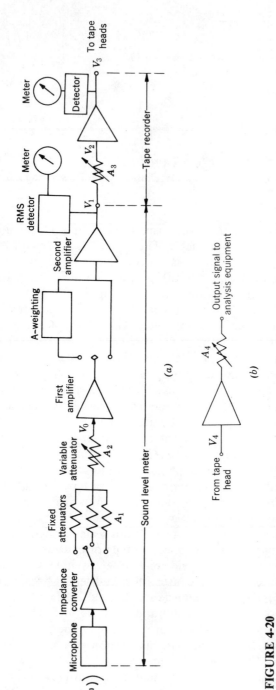

FIGURE 4-20

(*a*) Simplified functional diagram of a sound level meter and tape recorder input amplifier. (*b*) Tape recorder output amplifier.

116

downward to u dB for the sound level meter to register approximately $1/4$ to full scale. In other words, the sounds being recorded are approximately $x - u$ dB less than the calibration signal from the pistonphone. This value (u dB) must be permanently noted, usually with a voice track on the tape, for use upon playback. At the end of the tape (or recording session) the pistonphone is again used to put a calibration signal on the tape. This will indicate whether anything happened during recording to upset the calibration. If the two calibration signal levels differ substantially (> 1.0 dB) the recording should be discarded and a new one obtained.

For playback we refer to Figure 4-20b. The beginning of the tape containing the 10-20 sec of the calibration signal is played and the attenuator, A_4, is adjusted so that the appropriate levels exist in all the analyzing equipment that follow. This level corresponds to $y + u$ dB re 20 μPa. Notice that A_4 can be adjusted to any convenient setting on playback, as long as it is not changed during the course of the playback of the recorded sounds.

The Analysis of a Recorded Signal: An Example

The results of the preceding section are now illustrated. Consider a sound level meter that contains a microphone whose frequency response is given in Figure 4-17a. The output of the sound level meter is connected to a portable tape recorder. The sound-pressure level from the pistonphone is 123.8 dB re 20 μPa at 250 Hz. To put this calibration signal into the system the attenuator (meter range) of the sound level meter is set to 120 dB and the weighting network is switched to linear (or "C"-weighting). On some sound level meters this cannot be done. However, these are calibrated at 1000 Hz where the A-weighting is 0 dB. (Recall Figure 3-1.) Full scale on the sound level meter is 10 dB. The attenuator on the tape recorder is adjusted so that its meter reads -8 dB from its maximum allowable input level. In the notation of the section above we have that $x = 120$ dB, $y = 3.8$ dB, and $w = 8$ dB.

The levels of the sounds of interest are now ready to be recorded. It is found that the attenuator (A_1) on the sound level meter must be set to 80 dB so that the meter deflects, but never beyond the 10 dB mark (full scale) on the meter scale. This value is recorded for future use. From the preceding section, we have that $u = 80$ dB. With the attenuator on the sound level meter set at this level the sounds of interest are recorded. Before the tape is removed another calibration signal is placed on the tape.

The signals on the tape are then brought back to the laboratory and a $1/3$-octave analysis from 25 to 12,500 Hz is performed. It is desired that a statistical error of ± 1 dB be maintained. Using the procedure outlined in

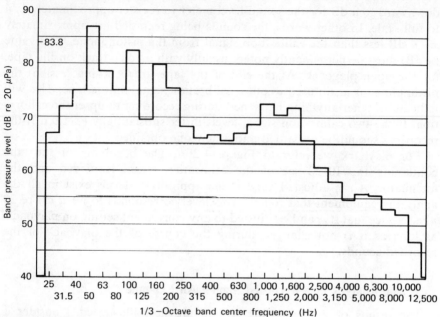

FIGURE 4-21
A typical 1/3-octave analysis from a calibrated tape.

Example 1–3 it is seen that this corresponds to an error of approximately ±12%. Hence from (4-4) the product of BT must equal 69.4. Since the narrowest bandwidth determines the longest averaging time required, we have, from (4-4) and Table 1-3 that for the 1/3-octave filter centered at 25 Hz, $B = 5.8$ Hz, and therefore, $T \geqslant 12$ sec.

The output of the analysis is presented on a graphic display device that records the amplitude on a logarithmic scale such that each division is 1 dB and the total range is 50 dB. See Figure 4-21. First the calibration signal is fed through the 1/3-octave filter having a center frequency of 250 Hz. The settings on the display device are adjusted so that its stylus deflects to 6.2 divisions (dB) below the top line on the paper. This level corresponds to 83.8 dB re 20 μPa. Hence the top of the chart paper corresponds to 90 dB and the bottom to 40 dB re 20 μPa.

REFERENCES

1. E. B. Magrab and D. S. Blomquist, *Measurement of Time-Varying Phenomena: Fundamentals and Applications*, Wiley, New York (1971).

2. "Octave, Half-Octave, and Third-Octave Band Filter Sets," ASA S1.11-1966 (R 1971), American National Standards Institute, New York.

3. "Measuring Microphones—Selected Reprints from *Technical Review*," Brüel & Kjaer, Naerum, Denmark (September 1972).

4. G. M. Sessler and J. E. West, "Electret Transducers: A Review," *J. Acoust. Soc. Am.*, Vol. 53, No. 6 (June 1973), pp. 1589-1600.

5. G. Rasmussen, "Measurement Microphones," INTER-NOISE 74 Proceedings, Washington, D. C. (September 1974), pp. 53-60.

6. "Specification for Sound Level Meters," ANSI S1.4-1971, American National Standards Institute, New York.

7. D. S. Blomquist, "An Experimental Investigation of Foam Windscreens," Paper No. G24y2, Proceedings of INTER-NOISE 73, Technical University of Denmark, Copenhagen (August 1973) pp. 589-593.

8. R. B. Wilkerson, "Sound Level Meter and Dosimeter Response to Unsteady Levels," INTER-NOISE 74 Proceedings, Washington, D. C. (September 1974), pp. 85-88.

5

NOISE SOURCES

5.1 INTRODUCTION

A large amount of work in recent years has been directed toward the development of empirical equations that can predict the sound-pressure or sound-power level from various noise sources. In some cases empirical equations have not been obtainable, but large amounts of data have been accumulated. It is the purpose of this chapter to present these data, which seem to fit naturally into several main areas: fan, pump, and compressor noise; duct noise; valve noise; machinery noise; and ground transportation noise. The results presented in the subsequent sections provide reasonably good estimations of the overall sound-power or sound-pressure levels of these more common sources of noise. However, these results should always be used with caution for the standard deviation, which is almost never indicated by the original investigators, may be large (>3 dB). Furthermore, the environment into which some of these sources might be placed may differ from those in which the measurements were performed and suitable adjustments should be made. This aspect is discussed in detail in Chapter 6.

5.2 NOISE FROM FANS, PUMPS, AND COMPRESSORS

Fan Noise—A General Discussion[1]

To minimize the necessity of reducing noise from fans with the installation of duct lining materials and duct bends the following factors should be considered:

1. The air distribution system should be designed for minimum resistance since fan sound generation, regardless of fan type, increases with static pressure.

2. Fans with relatively few blades (less than 15) tend to generate pure tones that may dominate the spectrum. These occur at the blade passage frequency and its harmonics. (The blade passage frequency is the product of the number of blades and the blade rotation speed in revolutions per second.) The intensity of these tones depends on resonances within the duct system and on fan design, as well as inlet flow distortions.

3. The fan should be selected to operate near its maximum efficiency point when handling the required air quantity and static pressure. Proper size is important in assuring a minimum of sound for any given type of fan. With regard to energy, sound generation and efficiency are not directly related because of the exceedingly small amount of energy required to produce sound. However, the same factors that reduce fan efficiency also increase sound.

If one were to compare both an oversized and undersized centrifugal fan with an optimum one the following results would be found. With only a 3% decrease in efficiency, the sound power of the oversized fan would be 3 dB higher than that of the optimum size fan. This increased sound generation is due primarily to separation of air flow along the blades.

An undersized fan, on the other hand, will operate at a higher speed than necessary. Its efficiency will be lower and sound-power levels will be substantially higher than for the optimum size fan. The primary causes of both lower efficiency and increased sound appear to be higher air-flow velocities, associated increased turbulence, and vortex shedding intensity.

4. Duct connections at both the fan inlet and outlet should be designed for uniform and straight air flow. Gusty and swirling inlet air flow particularly should be avoided. Variations from accepted application arrangements can severely degrade both the aerodynamic and acoustic performance of any fan type and invalidate manufacturers' ratings or other performance predictions.

Predictions of Centrifugal Fan Sound Power

The sound power generation of a given fan performing a given duty is best obtained from the manufacturer's actual test data taken under approved test conditions. If actual test data are not readily available, the octave band sound-power levels for a centrifugal fan can be estimated by the following procedure.[2] The method provides the sound level at any operating point, accounting for the change in sound level at various operating points through changes in static efficiency.

The sound-power level, L_w, of centrifugal ventilating fans with backward- and forward-curved blades is broken down into three components representing the various operating parameters:

$$L_w = A + B + T \qquad \text{dB re } 10^{-12} \text{ W} \qquad (5\text{-}1)$$

where A represents the air-flow volume and fan static pressure, B the fan static efficiency, and T the blade passage frequency tone. Equation 5-1 is valid to within ± 3 dB with a confidence of 95%.

The values for component A are obtained from Figure 5-1 at the given fan volume flow (m^3/sec) and static pressure (cm H$_2$O), regardless of fan type. The values of B are obtained from Figures 5-2 and 5-3 for the particular fan type using the static efficiency, η_s, at the actual point of operation. In these figures ϕ is the (dimensionless) flow coefficient, which is equal to a constant times the volume flow rate (m^3/sec) divided by the product of the cube of the fan rotor tip diameter (m) and the fan rotational speed (rev/sec). The quantity ϕ_{opt} is the flow coefficient at the peak static efficiency point of rating. These values are determined from data supplied by the fan manufacturer. The value of T is added only to the level in the

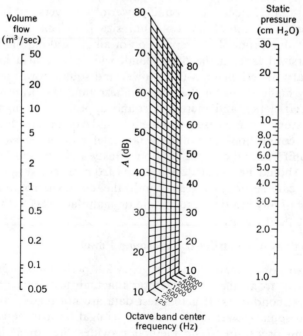

FIGURE 5-1
Nomograph for the determination of the parameter A in (5-1). (Reprinted with permission of the American Society of Heating, Refrigerating, and Air Conditioning Engineers, Inc. from ASHRAE TRANSACTIONS 1967, Volume 73, Part II Copyright © 1967.)

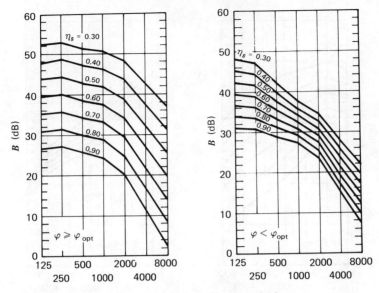

Octave band center frequency (Hz)

FIGURE 5-2
Graph of the parameter B in (5-1) for backward-curved blade fans. (Reprinted with permission of the American Society of Heating, Refrigerating, and Air Conditioning Engineers, Inc. from ASHRAE TRANSACTIONS 1967, Volume 73, Part II Copyright © 1967.)

octave band containing the blade passage frequency, f_B, which is determined from the relation

$$f_B = \frac{(N)(\text{rpm})}{60} \qquad \text{Hz} \qquad (5\text{-}2)$$

where N is the number of fan blades. In the absence of other information, a typical value of 5 dB for T may be assumed. If there is evidence that the particular fan to be used does not generate pure tones, T is zero.

For two or more identical fans operating in parallel, the combined sound-power output can be obtained with only one set of calculations by using the total system flow volume, the common static pressure, and the efficiency.

Another method for estimating centrifugal fan sound power is by the following expression:[3]

$$L_w = K + 10\log_{10} Q + 20\log_{10} p \qquad \text{dB re } 10^{-12} \text{ W} \qquad (5\text{-}3)$$

where Q is the volume flow rate (m³/sec), p is the static pressure (cm H_2O), and K, which is the specific sound-power level as a function of the fan type and the octave band center frequency, is given in Table 5-1.

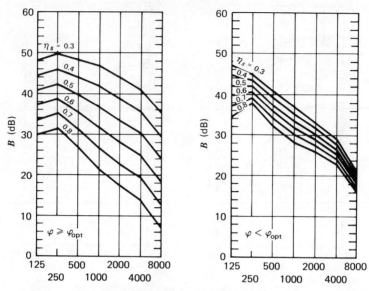

Octave band center frequency (Hz)

FIGURE 5-3
Graph of the parameter B in (5-1) for forward-curved blade fans. (Reprinted with permission of the American Society of Heating, Refrigerating, and Air Conditioning Engineers, Inc. from ASHRAE TRANSACTIONS 1967, Volume 73, Part II Copyright © 1967.)

TABLE 5-1
Parameter K in Equations 5-3[a]

	K (dB)								
	Octave band center frequency (Hz)								
Fan type	63	125	250	500	1000	2000	4000	8000	BPC[b] (dB)
Centrifugal									
Airfoil blade	60	60	59	57	56	51	43	35	3
Backward-curved blade	60	60	59	57	56	51	43	35	3
Forward-curved blade	65	63	63	59	53	49	46	40	2
Radial blade	73	70	68	68	63	58	55	54	5–8
Tubular	71	68	68	63	62	57	53	50	4–6
Vane axial	67	64	66	67	65	62	60	50	6–8
Tube axial	69	67	71	69	67	65	62	55	6–8
Propeller	76	73	74	72	70	70	68	56	5–7

[a]Reprinted, by permission, from Reference 3.
[b]This value, the blade passage correction, is added to the level in that octave band containing the blade passage frequency given by (5-2).

124

Example 5-1. A forward-curved fan with 45 blades has been selected to supply 4 m³/sec at a static pressure of 4.0 cm H₂O. The manufacturer's catalog indicates that 1200 rpm and 3.5 hp are required.

Solution: The fan static efficiency at this operating point is determined from the expression

$$\eta_s = \frac{(\text{cm H}_2\text{O})(\text{m}^3/\text{sec})}{7.612 \,(\text{hp})} = \frac{(4)(4)}{(7.612)(3.5)} = 0.600$$

It will be assumed that the fan is operating at its peak efficiency so that $\varphi = \varphi_{\text{opt}}$.

The blade passage frequency, obtained from (5-2), is

$$f_B = \frac{(1200)(45)}{60} = 900 \text{ Hz}$$

Thus, 5 dB is added to the octave band centered at 1000 Hz.

The results are summarized in Table 5-2.

TABLE 5-2
Tabulations of the Solution to Example 5-1 and 5-2

Octave band center frequency (Hz)	A Figure 5-1 (dB)	B Figure 5-3 (dB)	T Equation 5-2 (dB)	Example 5-1 Sound-power level, Equation 5-1 (dB re 10^{-12} W)	Example 5-2 Sound-power level, Equation 5-3 (dB re 10^{-12} W)
125	43	37	—	80	81
250	43	39	—	82	81
500	43	35	—	78	77
1000	44	32	5	81	73[a]
2000	44	28	—	72	67
4000	44	25	—	69	64
8000	45	18	—	63	58

[a] Includes the BPC.

Example 5-2. Repeat Example 5-1 using (5-3).

Solution: Equation 5-3 becomes

$$L_w = K + 10\log_{10}4 + 20\log_{10}4 = 18.1 + K \qquad \text{dB re } 10^{-12} \text{ W}$$

Using Table 5-1 the results are tabulated in the last column of Table 5-2. As can be seen there is reasonable agreement between the results of the two methods.

Noise from Induced Draft Fans[4]

The maximum sound-power level, L_w, at the discharge of induced draft fans at the blade passage frequency can be estimated from the empirical equation

$$L_w = 86 + 10\log_{10}\text{hp} + 10\log_{10}SP \qquad \text{dB re } 10^{-12} \text{ W} \qquad (5\text{-}4)$$

where hp is the rated horsepower of the fan and SP is the static pressure rating (centimeters of water). Typical values of hp range from 750 to 7500 and those of SP from 125 to 200 cm H_2O.

Example 5-3.   An induced draft fan is to deliver 60 m^3/sec at a static pressure head of 150 cm H_2O. The fan has 8 blades and requires 1750 hp at 1200 rpm. Determine an estimate of the sound-power level at the blade passage frequency.

Solution: From (5-2) it is found that the blade passage frequency is $f_B = 160$ Hz. At this frequency the estimated sound-power level, from (5-4), is found to be

$$L_w = 86 + 10\log_{10}1750 + 10\log_{10}150$$

$$= 140.2 \text{ dB re } 10^{-12} \text{ W}$$

Variations of Fan Noise with Size, Static Pressure, Speed, and Capacity[5]

The empirical relationships estimating the change in the overall sound-power level between the noise generated by a fan and its size S, static pressure P, speed V, and capacity Q, are as follows:

$$\Delta L_w = 70\log_{10}\frac{S_2}{S_1} + 50\log_{10}\frac{V_2}{V_1} \qquad \text{dB} \qquad (5\text{-}5a)$$

$$\Delta L_w = 20\log_{10}\frac{S_2}{S_1} + 25\log_{10}\frac{P_2}{P_1} \qquad \text{dB} \qquad (5\text{-}5b)$$

$$\Delta L_w = 10\log_{10}\frac{Q_2}{Q_1} + 20\log_{10}\frac{P_2}{P_1} \qquad \text{dB} \qquad (5\text{-}5c)$$

These relationships apply at a fixed point of rating of a fan. Fan size S refers to wheel diameter, housing height, or some dimension that is directly proportional to linear units.

Example 5-4.   Which is louder: a single-stage fan or a six-stage fan each having the same capacity and pressure.

Solution: Each stage of the six-stage fan operates at one-sixth of the pressure of the single-stage fan and one-sixth the capacity. Thus $Q_2/Q_1 = 6$ and $P_2/P_1 = 1/6$ and (5-5c) yields

$$\Delta L_w = 10\log_{10}6 - 20\log_{10}6 = -7.8 \text{ dB}$$

Thus the multistage fan generates almost 8 dB less overall sound power.

Example 5-5.   A fan having a wheel diameter of 1.0 m operates at 700 rpm. How much would the overall sound-power level increase if a fan having a wheel diameter of 1.3 m and a speed of 1000 rpm is substituted.

Solution: Using (5-5a), the increase in the overall sound-power level is found to be

$$\Delta L_w = 70\log_{10}\frac{1.3}{1.0} + 50\log_{10}\frac{1000}{700} = 15.7 \text{ dB}$$

Example 5-6.   Consider two fans, each of which operates at the same static pressure. One fan has a wheel diameter of 0.9 m and a capacity of 4.7 m^3/sec. The other fan has a wheel diameter of 1.2 m and a capacity of 8.5 m^3/sec. What is the change in overall sound-power level caused by the larger fan.

Solution: The change in level can be obtained from either (5-5b), which yields,

$$\Delta L_w = 20\log_{10}\frac{1.2}{0.9} + 25\log_{10}1 = 2.5 \text{ dB}$$

or from (5-5c), which yeilds,

$$\Delta L_w = 10\log_{10}\frac{8.5}{4.7} + 20\log_{10}1 = 2.6 \text{ dB}$$

Hydraulic Noise in Pumps

A theoretical and experimental investigation[6] of hydraulic noise produced at the discharge of volute and diffuser pumps led to the following relations for the overall sound-pressure level, L, which are true to within ±2 dB:

$$L = 20\log_{10}\frac{P_p}{S_0} + 94 \qquad \text{dB re 20 } \mu\text{Pa} \qquad (5\text{-}6)$$

where

$$P_p = \frac{Q(\Delta P)}{\Omega a_2^2 t_r} \qquad \text{Pa}$$

is called the power consumption and may have values from 10^3 to 10^5 and

$$S_0 = \frac{\Omega \sqrt{Q}}{(\Delta P/\rho)^{3/4}}$$

is called the dimensionless specific speed, which may have value from 500 to 9000. The quantity Q is the pump throughflow (m^3/sec), ρ is the density of the fluid (kg/m^3), Ω is the angular rotational speed (rad/sec), t_r is the thickness of the impeller blade at the outer radius (m), a_2 is the outer radius of the impeller (m), and ΔP is the pump pressure head (N/m^2). From (5-6) and the definition of P_p and S_0 it is seen that L decreases as the square of the pump speed Ω.

Airborne Noise from Pumps

Airborne sound-power levels generated by pumps with a rated speed of 1600 rpm or greater can be approximated by[7-9]

$$L_w = K_0 + 63 + 10\log_{10}\text{hp} \qquad \text{dB re } 10^{-12} \text{ W} \qquad (5\text{-}7)$$

where hp is the rated horsepower of the pump motor and K_0, which is a function of the type of pump and the octave band center frequency, is given in Table 5-3. If the rated speed is less than 1600 rpm subtract 5 dB from the above result.

TABLE 5-3
Parameter K_0 in Equation 5-7

Pump type	K_0 (dB) Octave band center frequency (Hz)								
	31.5	63	125	250	500	1000	2000	4000	8000
Centrifugal	25	25	26	26	27	29	26	23	18
Screw	30	30	31	31	32	34	31	28	23
Reciprocating	35	35	36	36	37	39	36	33	28

Example 5-7. Estimate the octave band sound-power level of a reciprocating pump that is operating at 2000 rpm and is rated at 10 hp.

Solution: Using (5-7) we have

$$L_w = K_0 + 63 + 10\log_{10}10 = K_0 + 73 \text{ dB re } 10^{-12} \text{ W}$$

From Table 5-3 we find that the octave band power levels in the nine octave bands from 31.5 to 8000 Hz are, respectively, 108, 108, 109, 109, 110, 112, 109, 106, and 101 dB re 10^{-12} W.

Centrifugal Compressors for Large Refrigerant Units

Data from the following two types of large centrifugal compressors have been obtained:[10] hermetically sealed and open motor-drive. The hermetically sealed refrigerator-cooled drive configurations are used in apartment buildings and appear to be the trend in liquid chilling equipment in the size range from 80 to 1300 tons refrigeration. Although open motor-drive equipment is still occasionally used in this range they are most often found in the size range above 1000 hp.

In the open motor-drive system motor noise is significant because the ratio of the physical size of the total unit to the motor horsepower is small, which requires the cooling air to circulate at relatively high velocities within a small space and results in considerable aerodynamic noise. In a hermetic system, the motor is cooled by the through-flow of refrigerant gas on the suction side of the compressor; any noise produced in the process becomes airborne only by structural radiation from the casing. Thus motor noise is rarely a problem. However, with the open motor-drive system motor noise may completely dominate if too much power is put into too small a frame.

Octave band noise levels from numerous hermetic and open motor-drive centrifugal compressors are shown in Figures 5-4 and 5-5, respectively.

Air Compressors[7]

For both reciprocating and centrifugal compressors within the range of 1–100 hp the sound-power level in each octave band can be approximated by

$$L_w = K_1 + 70 + 10 \log_{10} \text{hp} \qquad \text{dB re } 10^{-12} \text{ W} \qquad (5\text{-}8)$$

where hp is the rated horsepower of the drive motor and K_1, which is a function of the octave band center frequency, is given in Table 5-4.

TABLE 5-4
Parameter K_1 in Equation 5-8

	Octave band center frequency (Hz)								
	31.5	63	125	250	500	1000	2000	4000	8000
K_1 (dB)	18	14	12	8	7	11	12	9	4

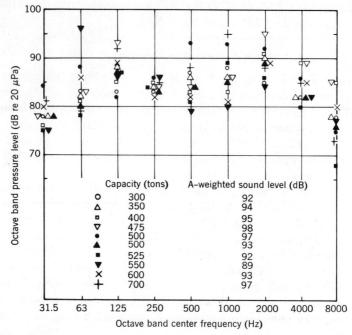

FIGURE 5-4

Typical hermetic centrifugal compressor octave band levels at full-load operating conditions and measured 0.9 m from the noisiest side. (Reprinted, by permission, from Reference 10.)

Example 5-8. Estimate the octave band power levels of an air compressor rated at 35 hp.

Solution: Using (5-8) gives

$$L_w = K_1 + 70 + 10\log_{10} 35 = 85.4 + K_1 \qquad \text{dB re } 10^{-12} \text{ W}$$

From Table 5-4 we find that the octave band power levels in the nine octave bands from 31.5 to 8000 Hz are, respectively, 103.4, 99.4, 97.4, 93.4, 92.4, 96.4, 97.4, 94.4, and 89.4 dB re 10^{-12} W.

5.3   FLOW NOISE IN DUCTS

Fans are a major, but not the only, source of sound to be considered in the design of the quiet duct systems. Air flow through elbows, dampers, branch take-offs, mixing units, sound traps, and other duct elements also produce sound. The sound-power levels in each octave band depend on the geometry of the device, the air turbulence, and the air flow rate.

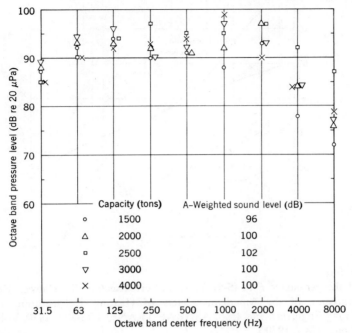

FIGURE 5-5
Typical open-drive centrifugal compressor octave band levels at full-load operating conditions and measured 0.9 m from the noisiest side. (Reprinted, by permission, from Reference 10.)

There is no limit to the number of geometries of duct elements used in industry. Attempts to develop a truly universal law for flow noise generation in duct elements have proven futile. On the other hand, it was demonstrated that useful estimates can be made for certain types of elements.

Elbows and branch take-offs are particularly important sound sources. Sound generated by duct fittings may be estimated with the following relation:[11]

$$L_w = F + G + 10\log_{10}f'_B - 1.5 \qquad \text{dB re } 10^{-12}\text{ W} \qquad (5\text{-}9)$$

where L_w is the octave band sound-power level, F is a spectrum function, G is a velocity function, and f'_B is the octave band center frequency.

Values for the spectrum function F are given in Figure 5-6 for square cross-section 90° miter elbows both with and without circular arc turning vanes, and in Figure 5-7 for one type of rectangular cross-section 90° miter elbows without turning vanes and with one central splitter. The spectrum functions in both figures are given in terms of the dimensionless Strouhal

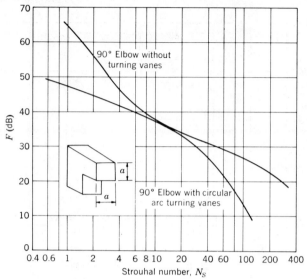

FIGURE 5-6
Graph of the parameter F in (5-9) for square cross-section 90° elbows. (Reprinted with permission of the American Society of Heating, Refrigerating, and Air Conditioning Engineers, Inc. from ASHRAE TRANSACTIONS 1970, Volume 76, Part II Copyright © 1970.)

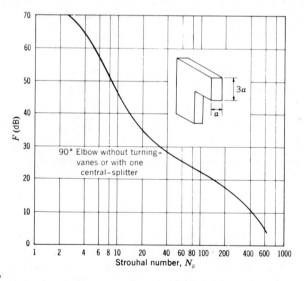

FIGURE 5-7
Graph of the parameter F in (5-9) for rectangular cross-section 90° elbows. (Reprinted with permission of the American Society of Heating, Refrigerating, and Air Conditioning Engineers, Inc. from ASHRAE TRANSACTIONS 1970, Volume 76, Part II Copyright © 1970.)

132

number:

$$N_s = \frac{f'_B D}{V} \qquad (5\text{-}10)$$

where V is the upstream air velocity (m/sec) and D is the duct diameter (m). For rectangular ducts $D = 1.128 \sqrt{A_0}$, where A_0 is the duct area (m^2).

Values for the velocity function G for 90° miter elbows needed in (5-9) may be obtained from Figure 5-8, which applies to both types of elbows. The G function is given in terms of the average duct velocity and the elbow cross-sectional area.

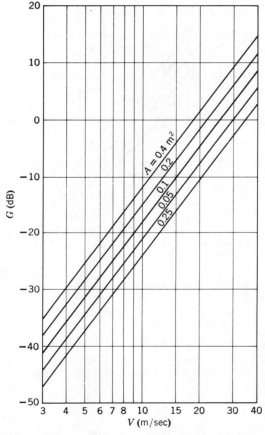

FIGURE 5-8
Graph of the parameter G in (5-9) for 90° elbows. (Reprinted with permission of the American Society of Heating, Refrigerating, and Air Conditioning Engineers, Inc. from ASHRAE TRANSACTIONS 1970, Volume 76, Part II Copyright © 1970.)

Aerodynamic sound produced by 90° branch tees has been correlated in a manner similar to that used for 90° elbows. Figure 5-9 shows the spectrum function F in terms of the cross-sectional area of the outlet duct and the dimensionless Strouhal number based on the upstream main duct diameter and the upstream duct velocity. For rectangular ducts, the equivalent diameter is $D = 1.128\sqrt{A_1}$ where A_1 is the upstream duct cross-sectional area. .

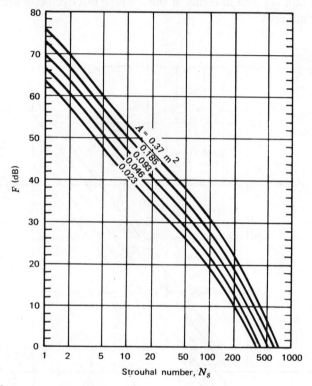

FIGURE 5-9

Graph of the parameter F in (5-9) for 90° branch take-off. (Reprinted with permission of the American Society of Heating, Refrigerating, and Air Conditioning Engineers, Inc. from ASHRAE TRANSACTIONS 1970, Volume 76, Part II Copyright © 1970.)

The velocity function G is given in Figure 5-10 in terms of the branch flow average velocity, V_3, and the downstream main duct velocity, V_2. Two functions are shown: one for a sharp-edge branch and one for a round-edge branch. It may be noted that for sufficiently high branch duct velocities, the velocity function is nearly independent of the downstream duct velocity; the converse is also true. Comparison of the two halves of Figure 5-10 shows that the round-edge branch has less velocity dependence, lower sound generation at high velocities, and higher sound generation at low velocities.

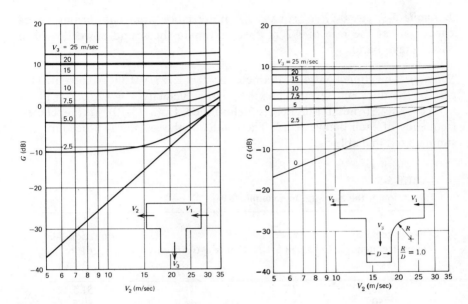

FIGURE 5-10
Graph of the parameter G in (5-9) for two types of 90° branch take-offs. (Reprinted with permission of the American Society of Heating, Refrigerating, and Air Conditioning Engineers, Inc. from ASHRAE TRANSACTIONS 1970, Volume 76, Part II Copyright © 1970.)

This procedure provides the sound-power radiated into either the branch duct or into the downstream duct by using the appropriate outlet duct area, A, in reading the spectrum function F from Figure 5-9.

Flow-generated noise also occurs in duct silencers (see p. 228) in which gas at a high velocity flows. In this case it is possible for a duct silencer with high sound attenuation to have the flow-generated noise exceed the noise attributable to the source. This problem of duct-silencer self-generated noise has been investigated by Ver[12] who gives the following expression for the approximate sound-power level at the exit plane of the silencer, and which is valid for any octave band from 125 to 8000 Hz:

$$L_{oct} \approx 140.2 + 55 \log_{10} V + 10 \log_{10} A - 45 \log_{10} P$$

$$- 20 \log_{10} T \quad \text{dB re } 10^{-12} \text{ W} \tag{5-11}$$

where V is the face velocity (m/sec), A is the face area (m²), P is the percentage cross-sectional open area, and T is the gas temperature (°K).

Example 5-9. A 0.37 m² square 90° miter elbow that has turning vanes carries air at the rate of 4.2 m³/sec. Estimate the sound-power levels in each octave band.

Solution: The equivalent duct diameter is: $D = 1.128 \sqrt{0.37} = 0.686$ m. The area of the duct is 0.37 m². Thus the average velocity $V = 4.2/0.37 = 11.35$ m/sec. The Strouhal number is determined from (5-10) at the center frequency of each octave band. The results of the graphical procedure are summarized in Table 5-5.

TABLE 5-5
Tabulation of the Solution to Example 5-9

Octave band center frequency (Hz)	N_s Equation 5-10	F Figure 5-6 (dB)	G Figure 5-8 (dB)	L_w Equation 5-9 (dB re 10^{-12} W)
125	7.7	39	−7	51.5
250	15.3	35	−7	50.5
500	30.6	30	−7	48.5
1000	61.3	21	−7	42.5
2000	122.5	10	−7	34.5
4000	245.0	−10[a]	−7	17.5
8000	490.0	−30[a]	−7	0

[a]Extrapolated.

Example 5-10. A 90° square branch take-off has a main duct with a 0.6×0.6 m cross-section both upstream and downstream. The branch duct is 0.3×0.46 m. The upstream main duct flow rate is 4.2 m³/sec and the branch flow rate is 0.566 m³/sec. Estimate the sound-power level in both the branch take-off and the downstream main duct.

Solution: The equivalent duct diameter of the main downstream duct is again 0.686 m. The area of the main downstream duct is 0.36 m² and that of the branch take-off is 0.14 m². Thus the downstream main duct velocity is $(4.2-0.57)/0.36 = 10.1$ m/sec, whereas that of the branch take-off is $0.57/0.14 = 4.1$ m/sec. The results of the graphical procedure are summarized in Table 5-6.

TABLE 5-6
Tabulation of the Solution to Example 5-10

Octave band center frequency (Hz)	N_s Equation 5-10	F Figure 5-9 (dB)	G Figure 5-10 (dB)	L_w Equation 5-9 (dB re 10^{-12} W)
		Branch take-off		
125	7.7	51	−6	64.5
250	15.3	45	−6	61.5
500	30.6	39	−6	58.5
1000	61.3	32	−6	54.5
2000	122.5	24	−6	49.5
4000	245.0	15	−6	43.5
8000	490.0	3	−6	34.5
		Downstream main duct		
125	7.7	56	−6	69.5
250	15.3	49	−6	65.5
500	30.6	44	−6	63.5
1000	61.3	36	−6	58.5
2000	122.5	28	−6	53.5
4000	245.0	19	−6	47.5
8000	490.0	7	−6	38.5

Example 5-11.   A duct silencer has a total cross-sectional area of 0.4 m² and an open area of 0.2 m². If the air flowing through the silencer is traveling at 12 m/sec with a temperature of 80°C, estimate the octave band sound power levels.

Solution: Using (5-11) and the fact that the percentage open area is 50%, we have

$$L_{oct} \approx 140.2 + 55\log_{10}12 + 10\log_{10}0.4 - 45\log_{10}50 - 20\log_{10}353$$

$$\approx 68.2 \text{ dB re } 10^{-12} \text{ W}$$

5.4   GRILLE NOISE

The overall sound power level, L_w, from air-conditioning diffusers can be estimated from the relation[13]

$$L_w = 10 + 10\log_{10} S + 30\log_{10}\xi + 60\log_{10} u \qquad \text{dB re } 10^{-12} \text{ W} \qquad (5\text{-}12)$$

where S is the area of the duct cross-section prior to the diffuser (m²), u is the mean flow speed in the duct prior to the grid (m/sec), and ξ is the normalized pressure drop coefficient given by

$$\xi = \frac{\Delta P}{0.5\rho u^2} \qquad (5\text{-}13)$$

where ρ is the density of air (kg/m³) and ΔP is the pressure drop across the diffuser (N/m²). Typical diffuser configurations and associated values of ξ are given in Figure 5-11.

Noise spectra of various configurations do not exhibit identical shapes even when normalized to similar flow speeds and exhaust areas. Minor construction differences may emphasize different frequency regimes. Some diffusers will, if improperly designed, radiate discrete frequency noise. In practical noise control problems, however, a general spectrum shape can be used that fits most diffuser noise spectra, as shown in Figure 5-12. A tolerance about each curve of ±5 dB should be employed to account for the uncertainty of the procedure. Using Figure 5-12 the sound-power level in each 1/3-octave band is obtained from the expression

$$L'_w = L_N + 10\log_{10} S + 30\log_{10}\xi \qquad \text{dB re } 10^{-12} \text{ W} \qquad (5\text{-}14)$$

It is seen from (5-14) that the spectrum shape is unaltered, only shifted in magnitude to obtain the sound-power level.

Example 5-12. Determine an estimate of the sound-power-level spectrum that results from air flow through a diffuser at a velocity of 4.0 m/sec. The pressure drop at this speed is 75 N/m² and the diffuser area is 0.133 m².

Solution: Assuming air has a density of 1.2 kg/m³, (5-13) yields

$$\xi = \frac{75}{(0.5)(1.2)(4)^2} = 7.8$$

From (5-12) we obtain

$$L'_w = L_N + 10\log_{10} 0.133 + 30\log_{10} 7.8 = L_N + 18 \qquad \text{dB re } 10^{-12} \text{ W}$$

The spectrum curve is that given by the curve of Figure 5-12 corresponding to $u = 4.0$ m/sec. This curve is redrawn in Figure 5-13 with the vertical axis now shifted upwards by 18 dB.

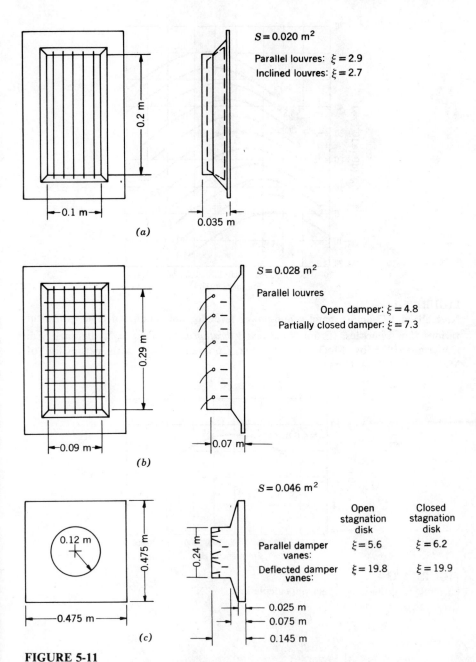

$S = 0.020\ \text{m}^2$

Parallel louvres: $\xi = 2.9$
Inclined louvres: $\xi = 2.7$

0.2 m

0.1 m

0.035 m

(a)

$S = 0.028\ \text{m}^2$

Parallel louvres

Open damper: $\xi = 4.8$
Partially closed damper: $\xi = 7.3$

0.29 m

0.09 m

0.07 m

(b)

$S = 0.046\ \text{m}^2$

	Open stagnation disk	Closed stagnation disk
Parallel damper vanes:	$\xi = 5.6$	$\xi = 6.2$
Deflected damper vanes:	$\xi = 19.8$	$\xi = 19.9$

0.12 m

0.475 m

0.24 m

0.475 m

0.025 m
0.075 m
0.145 m

(c)

FIGURE 5-11

Various duct grille configurations and their pressure drop coefficients. (From *Noise and Vibration Control* by L. L. Beranek, Ed. Copyright 1971 by McGraw-Hill Book Company. Used with permission of McGraw-Hill Book Company.)

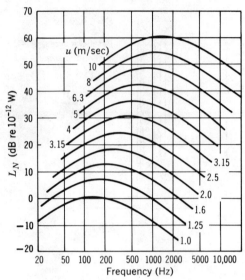

FIGURE 5-12
Normalized 1/3-octave band sound-power level for noise radiated from grilles for various flow velocities. (From *Noise and Vibration Control* by L. L. Beranek, Ed. Copyright 1971 by McGraw-Hill Book Company. Used with permission of McGraw-Hill Book Company.)

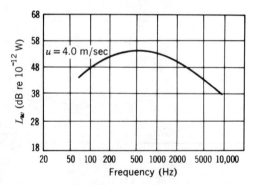

FIGURE 5-13
Example of grille-noise spectrum calculation.

5.5   VALVE NOISE

Introduction

Control valves, pressure regulators, and throttling devices are an integral part of many industrial plants. They are the pressure and flow controlling elements in all gas and liquid handling installations. Sound is generated by these devices when the energy contained in the compressible fluid is converted into sonic streams at the valves orifice. The velocity of these jets are in turn, slowed to the mean stream velocity by the turbulent mixing mechanism between the jet and the cavity gas in the space aft of and some distance downstream from the orifice. The noise of valves is caused by this confined jet mixing aft of the orifice and the subsequent disturbing effect on the surrounding fluid.

To date there are several empirically developed relations equating the sound-pressure or sound-power level from a valve as a function of various flow, valve, and pipe characteristics. Because of their empirical nature, these methods are discussed separately. For a critique of several of the methods discussed, see Reethof and Karvelis.[14] In addition, some fundamental studies on valve noise have recently been performed.[15, 16]

Nakano's Results[17]

Nakano obtained the following empirical relationship for the sound-power level L_w of noise emitted by a valve outlet into air:

$$L_w = A + B \log_{10}(GTF) \qquad \text{dB re } 10^{-12} \text{ W} \qquad (5\text{-}15)$$

where T is the total temperature in the inlet pipe in °K, G is the flow weight of the gas through the valve in kg/sec, and $F = 1 - (P_2/P_1)^{[(\gamma-1)/\gamma]}$, ($P_1$ is the total pressure at the valve inlet, P_2 is atmospheric pressure, and γ is the ratio of the specific heats of the gas). The constants A and B depend on the type of valve and are given in Table 5-7.

TABLE 5-7
Values of A and B in Equation 5-15

Valve type	A	B
Globe	90	10.0
Gate	83	15.6
Diaphragm	72	19.7
Ball	97	12.8
Angle	82	13.1

Example 5-13. Air at 120°C is flowing through a gate valve into the atmosphere at a rate of 0.04 kg/sec. The ratio of the inlet pressure of the valve to the atmospheric pressure is 4.0. What is the estimated overall sound-power level from the exhaust of this valve.

Solution: The value of F is

$$F = 1 - (0.25)^{[(1.4 - 1)/1.4]} = 0.327$$

From Table 5-7, $A = 83$ and $B = 15.6$. Using (5-15) the overall sound-power level is

$$L_w = 83 + 15.6 \log_{10}[(0.04)(0.327)(393)]$$

$$L_w = 94.1 \text{ dB re } 10^{-12} \text{ W}$$

Fisher Controls Company Method[18–21]

Fisher's procedure is an empirical method of predicting the total sound-pressure level in an "acoustically isolated chamber." These sound-pressure levels are at a position 1.2 m downstream from the valve and 0.74 m from the pipe surface. The relation they obtained is

$$L = (SPL)_{\Delta P} + (\Delta SPL)_{C_g} + (\Delta SPL)_{\Delta P/P_1} + (\Delta SPL)_K \qquad \text{dB re } 20 \text{ } \mu\text{Pa}$$

$$(5\text{-}16)$$

where $(SPL)_{\Delta P}$ is the base sound-pressure level determined as a function of ΔP, the pressure differential or pressure drop across the valve; $(\Delta SPL)_{C_g}$ is the correction for the variation in the critical flow coefficient C_g, which is experimentally determined for each valve and available from the valve manufacturer; $(\Delta SPL)_{\Delta P/P_1}$ is the correction for valve style and pressure drop ratio for specific valve sizes; and $(\Delta SPL)_K$ is the correction for the pipe wall attenuation as a function of pipe diameter and schedule number.

Since (5-16) is applicable only to Fisher valves the graphs describing the various terms in it will not be given here. They are presented elsewhere.[22] It has been pointed out by Reethof and Karvelis,[14] however, that there are several deficiencies implicit in the use of (5-16). First, there is no spectral data and, therefore, the levels cannot be converted to A-weighted levels. Second, there was no anechoic termination of the test pipe so that some of the data may have been influenced by standing waves. Third, the standard deviation of the test results is not given. Fourth, directional characteristics of the sound emitted from the pipe were not investigated.

Masoneilan Corporation Method[23, 24]

This approach is based on an analytical attempt to express the conversion of mechanical power in a valve to acoustic power. The method is primarily limited to sonic or choked flow conditions at the throat of the valve. The empirical formula they developed gives the A-weighted sound-pressure level 0.9 m away from the throttling control valve operating under choked flow conditions in an "acoustically soft" room. Their relationship is given by

$$L_A = 10\log_{10}\eta C_v C_f P_1 P_2 - TL - S_g + 110 \qquad \text{dB re } 20 \ \mu\text{Pa} \qquad (5\text{-}17)$$

where η is the acoustic efficiency (which is given as a function of valve pressure ratios and critical flow factors), C_v is the valve flow coefficient, C_f is the critical flow coefficient (which implicitly contains the valve geometry), P_1 is the static upstream absolute pressure, P_2 is the static downstream absolute pressure, TL is the transmission loss through the pipe (dB), and S_g is the gas property correction factor (dB). These results are only valid for Masoneilan's valves.

Equation 5-17 is based on the assumption that noise generated by turbulence in a free field is applicable to the case of confined jet mixing. The criticisms expressed for Fisher valves also apply for these valves. Another criticism is that no attention was given to the formation of shock cells downstream of the choked cross-section for overexpanded conditions. These cells involve sharp pressure discontinuities which in turn amplify the turbulence-generated pressure discontinuities.

Furthermore, (5-17) assumes that the sound power and sound pressure are simply related following the criteria for spherical radiation. It is well-known that a free jet does not radiate spherically and there is no reason to believe that a confined jet and associated piping will radiate spherically. Finally, it is also mentioned that the transmission loss term in (5-17) was based on that for a flat plate at the frequency associated with the peak value of the spectral distribution of the free jet, which may not have any relation to the transmission loss through a pipe wall.

5.6 MACHINERY NOISE

Electric Motors

The octave band sound-power level of the airborne noise generated by electric motors or motor-generator sets with horsepower ratings between 1 and 300 is given by:[7]

$$L_w = K_2 + 20 \log_{10} hp + 15 \log_{10} N - 7 \qquad dB \text{ re } 10^{-12} \text{ W} \qquad (5\text{-}18)$$

where hp is the rated horsepower of the unit, N its rated speed (rpm), and K_2, which is a function of the octave band center frequency, is given in Table 5-8.

TABLE 5-8
Values of K_2 in Equation 5-18

	Octave band center frequencies (Hz)								
	31.5	63	125	250	500	1000	2000	4000	8000
K_2 (dB)	7	9	13	15	16	16	14	7	0

To determine the noise caused by the magnetic fields of ac and dc electric machines see King.[25] For estimates of the octave band power levels for megawatt electric power generators see Reference 26.

Example 5-14. Estimate the octave band power levels from a 100 hp electric motor with a rated speed of 2600 rpm.

Solution: Equation 5-18 becomes

$$L_w = K_2 + 20 \log_{10} 100 + 15 \log_{10} 2600 - 7 = 84.2 + K_2 \qquad dB \text{ re } 10^{-12} \text{ W}$$

Using Table 5-8 it is found that the sound-power levels in the nine octave bands from 31.5 to 8000 Hz are, respectively, 91, 93, 97, 99, 100, 100, 98, 91, and 84 dB re 10^{-12} W.

Gear Noise

Mitchell[27] has provided guidelines, given in Table 5-9 and Figures 5-14 to 5-20, that indicate the change in the overall sound-pressure or sound-power level caused by a modification of an existing gear design. With regard to the table and figures, the potential reductions are not necessarily additive if multiple changes are made. Also, the AGMA (American Gear Manufacturers Association) number indicates the AGMA gear classification.*

*"AGMA Gear Classification Manual," AGMA 390.02 (September 1964).

TABLE 5-9
Noise Reduction in Gears as a Function of Various Design Parameters[a]

Design parameter	Range of possible reduction (dB)	Remarks
Profile error	0–5	Normal manufacturing
	5–10	Ultraprecision gears
Profile roughness	3–7	Full range of standard manufacturing techniques
Tooth spacing error	3–5	—
Tooth alignment error	0–8	—
Speed	$\sim 20\log_{10}(V/V_{ref})$	(V = Velocity)
Load	$\sim 20\log_{10}(L/L_{ref})$	(L = Load)
Power	$\sim 20\log_{10}(LV/L_{ref}V_{ref})$	
Pitch	—	Finer→quieter
Contact ratio	0–7	Largest best, but if small contact ratios are necessary use 2.0
Angle of approach and recess	—	Approach forces→higher; therefore, smaller approach angle→quieter
Pressure angle	—	Lower pressure angle→quieter
Helix angle	2–4	For changes from spur to helical
Tooth face width	0	No effect for equivalent specific loads
Gear tooth backlash	0–14	If excessive backlash
	3–5	If too little backlash
Air ejection effects	6–10	25 m/sec or more
Tooth phasing	—	Not practical
Planetary system phasing	5–11	Practical
Gear housing	6–10	If resonant
Gear damping	0–5	If resonant or needs isolation
Bearing	0–4	Adds damping; some types may stiffen structure
Bearing installation	0–2	Can increase life and eliminate some frequencies
Lubrication	0–2	Filled gearbox quietest

[a] Reprinted, by permission, from Reference 27.

Attia[28] has provided similar information about circular-arc gear teeth for gears with diametral pitches of 6 and 10. Some of his conclusions are:

(1) The total average noise-pressure level of gears of circular-arc tooth profile is generally higher than the noise of involute helical gears running under the same conditions and of similar parameters.

(2) Gear noise-pressure level increases rapidly with increase of speed until it reaches a maximum value. The effect of load in increasing gear noise is greater at high speeds.

Gear Noise Classification

Class A:	Noise behavior cannot be reliably obtained even with high quality production techniques. Additional sound absorption, vibration damping, vibration isolation, structural reinforcement are often required.
Class B:	Result of extremely high manufacturing accuracy and quality control.
Class C:	High manufacturing accuracy.
Class D:	Normal manufacturing quality.
Class E:	Gear drives with high noise levels that are easily corrected by increasing the manufacturing quality.

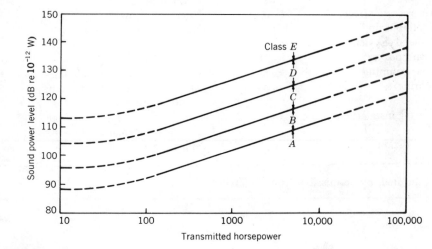

FIGURE 5-14

Noise-quality classification for gear systems. (Reprinted, by permission, from Reference 27.)

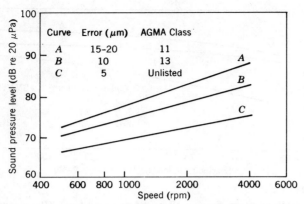

FIGURE 5-15
Influence of involute profile error on the overall sound-pressure level. (Reprinted, by permission, from Reference 27.)

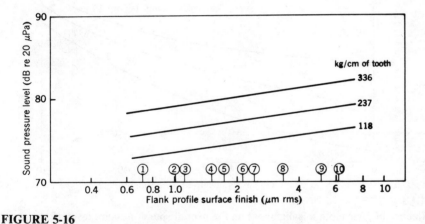

FIGURE 5-16
Influence of flank profile surface finish on the overall sound-pressure level for gears rotating at 1500 rpm. (Reprinted, by permission, from Reference 27.) (1) Smooth ground (fine ground); (2) milled and shaved; (3) smooth ground (coarse ground); (4) cross ground (electropolished); (5) cross ground (double-smooth ground); (6) cross ground (smooth ground); (7) cross ground (semismooth ground); (8) milled and lapped; (9) milled (soft); and (10) milled (heat-treated).

147

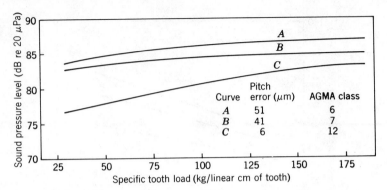

FIGURE 5-17
Influence of pitch error on the overall sound-pressure level. (Reprinted, by permission, from Reference 27.)

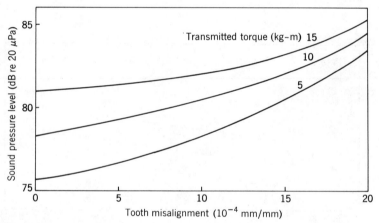

FIGURE 5-18
Influence of gear tooth misalignment on the overall sound-pressure level. (Reprinted, by permission, from Reference 27.)

(3) The number of teeth in contact between a pair of gears is an important factor in noise generation. A consistent abrupt rise in noise level has been recorded as the contact ratio approaches a whole number. For quieter gear running the designer is advised to choose higher diametral pitches and to avoid the choice of gear facewidths that are multiples of the axial pitches.

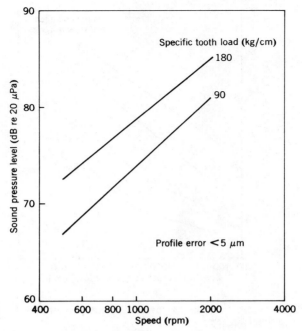

FIGURE 5-19
Influence of gear rotational speed on the overall sound-pressure level. (Reprinted, by permission, from Reference 27.)

Excitation Frequencies of Gears and Bearings

Consider a gear assembly consisting of two meshing gears. The drive gear has N_1 teeth and a shaft rotation of S_1 (rpm). The driven gear has N_2 teeth and a shaft rotation $S_2 = S_1 N_1 / N_2$ (rpm). The drive and driven shaft frequencies are, respectively,

$$f_1 = \frac{S_1}{60} \qquad \text{Hz} \tag{5-19}$$

and

$$f_2 = \frac{S_2}{60} = \frac{S_1 N_1}{60 N_2} \qquad \text{Hz} \tag{5-20}$$

The fundamental gear mesh frequency is

$$f_{gm} = \frac{S_1 N_1}{60} = f_1 N_1 \qquad \text{Hz} \tag{5-21}$$

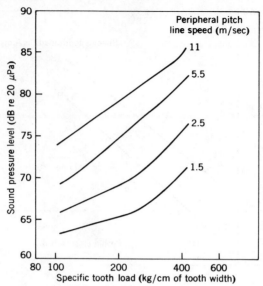

FIGURE 5-20
Influence of gear tooth load on the overall sound pressure-level. (Reprinted, by permission, from Reference 27.)

It is also possible, when the pitch diameter of the gear tooth is not exact, for the fundamental gear mesh frequency to generate what are known as upper and lower sideband frequencies. The frequencies are denoted f_{nU} and f_{nL}, respectively, where $n = 1, 2, 3, \ldots$. These frequencies are determined from the following simple relations:

$$f_{nU} = f_{gm} + nf_1 \quad \text{Hz}$$
$$f_{nL} = f_{gm} - nf_1 \quad \text{Hz} \qquad n = 1, 2, 3, \ldots \qquad (5\text{-}22)$$

The shafts to which the gears are attached are supported by bearings. The bearings themselves can, under certain circumstances, introduce noise levels at discrete frequencies due to imperfections in the bearings or their containers, called the inner and outer races. Consider the bearing shown in Figure 5-21. The inner race of the bearing is rigidly attached to the shaft and has a diameter a. It rotates at a speed of N_s (rpm). The balls (or rollers) have a diameter b and have a rotation speed of N_B (rpm) about their own axis. The rotation speed of the balls or rollers about the inner race, called the cage rotation speed, is N_c (rpm). There are a total of M balls or rollers. If it is assumed that there is no slippage between the balls or rollers and the races, the possible vibration frequencies that the bearing

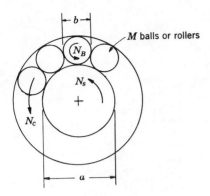

FIGURE 5-21
Typical roller or ball bearing geometry.

can produce are as follows:[25] The cage rotation speed is

$$N_c = N_s \frac{a}{2(a+b)} \qquad \text{rpm} \qquad (5\text{-}23)$$

The rotation speed of the balls or rollers about their own axis is

$$N_B = N_s \left(\frac{a}{2b} \right) \qquad \text{rpm} \qquad (5\text{-}24)$$

The frequency of impacts of the balls or rollers against an imperfection on the outer race is

$$f_o = \frac{aMN_s}{120(a+b)} \qquad \text{Hz} \qquad (5\text{-}25)$$

The frequency of impacts of balls or rollers against an imperfection on the inner race is

$$f_i = \frac{MN_s}{60} \left[1 - \frac{a}{2(a+b)} \right] \qquad \text{Hz} \qquad (5\text{-}26)$$

Example 5-15. Consider the gear and bearing system shown in Figure 5-22. The drive shaft rotates at 4000 rpm and has attached to it a gear with 29 teeth. The driven shaft has a gear with 62 teeth. The bearing contains 18 elements with a diameter of 15 mm. The diameter of the inner race is 75 mm. Determine the possible vibration frequencies up to and including the third sidebands.

Solution: From (5-19) and (5-20) the gear rotation frequencies are

$$f_1 = \frac{4000}{60} = 66.7 \text{ Hz} \qquad\qquad f_2 = \frac{(4000)(29)}{(60)(62)} = 31.2 \text{ Hz}$$

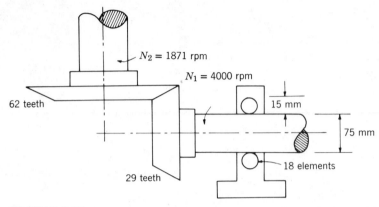

FIGURE 5-22
Gear and bearing geometry for Example 5-15.

Using (5-21) the gear mesh frequency is

$$f_{gm} = (66.7)(29) = 1933.3 \text{ Hz}$$

The upper and lower sideband frequencies are determined from (5-22). Thus

$$f_{nU} = 1933.3 + 66.7n \quad \text{Hz}$$
$$n = 1, 2, 3$$
$$f_{nL} = 1933.3 - 66.7n \quad \text{Hz}$$

The above results plus those that follow are summarized in Table 5-10. The frequencies due to a rough spot in the outer and inner race are determined from (5-25) and (5-26), respectively. Thus

$$f_o = \frac{(75)(18)(4000)}{(120)(75 + 13)} = 511.4 \text{ Hz}$$

$$f_i = \frac{(18)(4000)}{60} \left[1 - \frac{75}{2(75 + 13)} \right] = 688.6 \text{ Hz}$$

Noise From Home Appliances

Representative A-weighted levels have been obtained[29] for a wide variety of home appliances. These are summarized in Table 5-11. For examples of typical 1/3-octave spectra for these devices see Reference 29.

TABLE 5-10
Summary of Possible Frequencies Generated by System Shown in Figure 5-22

Component description	Frequency (Hz)
Rotation of driven shaft	31.2
Rotation of drive shaft	66.7
Rough spot on outer race	511.4
Rough spot on inner race	688.6
3rd Lower sideband	1733.2
2nd Lower sideband	1799.9
1st Lower sideband	1866.6
Gear mesh	1933.3
1st Upper sideband	2000.0
2nd Upper sideband	2066.7
3rd Upper sideband	2133.4

TABLE 5-11
A-Weighted Sound Levels at 0.9 m for Various Home Appliances[a]

Appliance	A-Weighted levels	
	Range (dB)	Mean (dB)
Refrigerator	35–52	42
Fan	38–69	57
Clothes dryer	51–66	58
Air conditioner	50–67	58
Electric shaver	47–69	60
Hair dryer	59–65	61
Clothes washer	47–72	62
Water closet	50–72	63
Dishwasher	54–72	65
Electric can opener	54–76	66
Food mixer	49–79	68
Vacuum cleaner	62–85	72
Food blender	62–88	75
Food waste disposer	67–93	78
Home shop tools	72–97	83

[a]From Reference 29.

Noise From Building and Construction Equipment

A-Weighted levels have been obtained[29] for a wide variety of typical building and construction equipment. A summary of the typical ranges of the A-weighted levels of building equipment is given in Table 5-12. A summary of the typical ranges of A-weighted levels for construction equipment is given in Table 5-13. Some limited data on farm equipment can be found in Reference 30. These references also contain some representative 1/3-octave band spectra.

Noise from Gas Turbines[7,8,26]

The noise from a gas turbine comes from three sources: the casing, the intake, and the exhaust. The sound-power level from the casing is estimated from the expression

$$L_C = K_C + 5\log_{10}hp + 92 \qquad dB \text{ re } 10^{-12} \text{ W} \qquad (5-27)$$

where K_C, which is a function of the octave band center frequency, is given in Table 5-14, and hp is the rated horsepower of the turbine.

TABLE 5-12
A-Weighted Sound Levels at 0.9 m for Typical Building Equipment [a]

Equipment	Range of A-weighted levels (dB)
Fluorescent lamp ballast	20–50
Fan coil units	25–55
Diffusers, grilles, and registers	20–60
Induction units	25–60
Dehumidifiers	40–70
Humidifiers	20–70
Mixing boxes, terminal reheat units, etc.	25–80
Unit heaters	45–80
Transformers	70–80
Boilers	55–90
Rooftop air-conditioning units	70–90
Pumps	45–92
Steam valves	60–92
Air-cooled condensers	80–95
Pneumatic transport systems	60–100
Air compressors	75–105
Cooling towers	85–110

[a] From Reference 29.

TABLE 5-13

A-Weighted Levels at 15 m from Typical Construction Equipment

Equipment	Range of A-weighted levels (dB)
Earth movers	
Front loaders	72–84
Backhoes	72–93
Tractors	76–96
Scrapers, graders	80–93
Pavers	86–88
Trucks	82–94
Material handlers	
Concrete mixers	75–88
Concrete pumps	81–83
Cranes (movable)	75–86
Cranes (derrick)	86–88
Stationary	
Pumps	69–71
Generators	71–82
Compressors	74–86
Impact	
Pneumatic wrenches	83–88
Jack hammers and rock drills	81–98
Pile drivers (peaks)	95–105

TABLE 5-14

Values of K_E and K_C in Equations 5-27 and 5-28

	Octave band center frequency (Hz)								
	31.5	63	125	250	500	1000	2000	4000	8000
K_C (dB)	1	4	6	7	7	7	7	7	7
K_E (dB)	22	22	22	22	22	20	16	11	4

The sound-power level from the exhaust is estimated from the expression

$$L_E = K_E + C_1 + 10\log_{10}\text{hp} + 73 \qquad \text{dB re } 10^{-12}\text{ W} \qquad (5\text{-}28)$$

where K_E, which is a function of the octave band center frequency, is also given in Table 5-14. The correction term C_1 is described subsequently.

The sound-power level from the turbine intake is estimated from the expression

$$L_I = K_I + C_2 + 15\log_{10}\text{hp} + 57 \qquad \text{dB re } 10^{-12}\text{ W} \qquad (5\text{-}29)$$

where K_I is an adjustment to account for the turbine compressor rotation rate (R_R), the blade rate (B_R), and the harmonics and subharmonics of these frequencies. For each of the frequencies calculated, the appropriate values of K_I are added to those octave bands in which these frequencies lie. (Recall Table 1-3.) The rotation rate is given by $R_R = N_S/60$ (Hz), where N_S is the compressor shaft rotation speed (rpm). The blade rate is given by $B_R = N_1 R_R$ (Hz), where N_1 is the number of blades of the first stage of the compressor. If two or more corrections are made in any octave band they must first be combined using (1-24) or Figure 1-9 before being inserted into (5-29). The values of K_I' (where $K_I = K_I'$ if there is only one frequency in the octave band) are given in Table 5-15.

The correction terms C_1 and C_2 reflect the adjustments made to bring the values of L_I and L_E into the source room, and reflect the transmission loss properties of the duct wall (see Section 7.4) and the length of the duct itself. This latter correction, given by $10\log_{10}(4L_0/D)$ where L_0 is the duct length and D is the smallest dimension of the duct cross-section, is added

TABLE 5-15
Values of $K_I'^a$

Harmonic (Hz)	K_I' (dB)
$0.25R_R$	1
$0.5R_R$	2
R_R	3
$2.0R_R$	2
$0.125B_R$	3
$0.25B_R$	6
$0.5B_R$	12
B_R	18
$2.0B_R$	15
$4.0B_R$	12

^{a}See text for the use of K_I'

to the transmission loss value, which is a negative quantity. At the intake and exhaust discharge planes of the ducts $C_1 = C_2 = 0$, respectively.

Example 5-16. A gas turbine has a rated horsepower of 20,000. The compressor shaft rotates at 9500 rpm and its first stage contains 38 blades. The intake duct is 1.4 m in diameter and 12 m long. The exhaust duct is 0.3 m in diameter and 21 m long. For simplicity the transmission loss of the duct is assumed to be -35 dB in each octave band. Estimate the total octave band sound-power level in the source room.

Solution: The values of C_1 and C_2 are, respectively,

$$C_1 = -35 + 10\log_{10}\frac{84}{0.3} = -10.5 \text{ dB}$$

$$C_2 = -35 + 10\log_{10}\frac{48}{1.4} = -19.6 \text{ dB}$$

Equations 5-27 to 5-29 yield, respectively,

$$L_C = K_C + 5\log_{10}20{,}000 + 92 = K_C + 113.5 \qquad \text{dB re } 10^{-12} \text{ W}$$

$$L_E = K_E + 10\log_{10}20{,}000 + 73 - 10.5 = K_E + 105.5 \qquad \text{dB re } 10^{-12} \text{ W}$$

$$L_I = K_I + 15\log_{10}20{,}000 + 57 - 19.6 = K_I + 101.9 \qquad \text{dB re } 10^{-12} \text{ W}$$

The compressor rotation rate is $R_R = 9500/60 = 158.3$ Hz and the blade rate is $B_R = (38)(158.3) = 6016.7$ Hz. Consequently the harmonics, corresponding to the order given in Table 5-15 are: 39.6, 79.2, 158.3, 316.6, 752.1, 1504.2, 3008.4, 6016.7, 12,033, and 24,066 Hz. Referring to Table 1-3 it is seen that the first frequency is in an octave band with the center frequency equal to 31.5, the second in the 63-Hz band, the third in the 125-Hz band, the fourth in the 250-Hz band, the fifth in the 1000-Hz band, the sixth in the 2000-Hz band, the seventh in the 4000-Hz band and the eighth in the 8000-Hz band. The last two frequencies are outside the 8000-Hz band.

Using Table 5-14 and 5-15 for the values of K_C, K_E, and $K_I = K_I'$ yields the results summarized in Table 5-16.

5.7 TRANSPORTATION NOISE: TRUCKS AND TRAINS

Diesel Engine Noise[31,32]

The A-weighted-sound-pressure level at 15 m for three types of diesel engines can be estimated from the following relations:

TABLE 5-16
Tabulation of the Solution to Example 5-16

Octave band center frequency (Hz)	L_C (dB)	L_E (dB)	L_I (dB)	L_{Total} (dB re 10^{-12} W)[a]
31.5	114.5	127.5	102.9	127.7
63.0	117.5	127.5	103.9	127.9
125	119.5	127.5	104.9	128.2
250	120.5	127.5	103.9	128.3
500	120.5	127.5	101.9	128.3
1000	120.5	125.5	104.9	126.7
2000	120.5	121.5	107.9	124.1
4000	120.5	116.5	113.9	122.6
8000	120.5	109.5	119.9	123.4

[a] From (1-24).

4-stroke, naturally aspirated

$$L_E = 30 \log_{10} N + 50 \log_{10} B - 70.7 \qquad \text{dB re } 20 \ \mu\text{Pa} \qquad (5\text{-}30)$$

4-stroke, turbocharged

$$L_E = 40 \log_{10} N + 50 \log_{10} B - 105.7 \qquad \text{dB re } 20 \ \mu\text{Pa} \qquad (5\text{-}31)$$

2-stroke

$$L_E = 40 \log_{10} N + 50 \log_{10} B - 96.7 \qquad \text{dB re } 20 \ \mu\text{Pa} \qquad (5\text{-}32)$$

where N is the engine speed (rpm) and B is the bore of the engine cylinder (cm).

Muffled Diesel Engine Exhaust Noise[31,32]

An estimate of the A-weighted-sound-pressure level at 15 m of muffled engine exhaust noise, based on a highly representative cross-section of numerous combinations of makes of mufflers and diesel engines, can be obtained from the following equation:

$$L_M = 10 \log_{10} \text{bhp} + 74.5 - C_0 \qquad \text{dB re } 20 \ \mu\text{Pa} \qquad (5\text{-}33)$$

where bhp is the brake horsepower of the engine and C_0 has the following values: 15 dB for a 2-stroke engine; 17.2 dB for a 4-stroke, naturally aspirated engine; and 16.7 dB for a 4-stroke, turbocharged engine.

Diesel Engine Fan Noise[31,32]

An estimate of the A-weighted-sound-pressure level at 15 m of the cooling fans on diesel engines is given by

$$L_F = 10\log_{10} b_f n_f + 30\log_{10}\left[(a_1 N d_f)^2 + (5.305\ V)^2\right]$$

$$-108.6 \qquad \text{dB re 20 } \mu\text{Pa} \qquad (5\text{-}34)$$

where b_f is the fan blade width (m), n_f is the number of fan blades, N is the engine speed (rpm), V is the vehicle speed (km/hr), d_f is the fan diameter (m), and a_1 equals 1.0 for engines whose displacement is less than 9800 cm^3 and 1.2 for engine displacements greater than or equal to 9800 cm^3.

Diesel Engine Intake Noise[31,32]

An estimate of the A-weighted sound-pressure level at 15 m due to three different configurations of intakes to the three types of engines described above can be obtained from the following expressions:

4-stroke, turbocharged

$$L_I = 63 + 5\log_{10} \text{bhp} - C_1 \qquad \text{dB re 20 } \mu\text{Pa} \qquad (5\text{-}35)$$

4-stroke, naturally aspirated

$$L_I = 81 - C_1 \qquad \text{dB re 20 } \mu\text{Pa} \qquad (5\text{-}36)$$

2-stroke

$$L_I = 83 - C_2 \qquad \text{dB re 20 } \mu\text{Pa} \qquad (5\text{-}37)$$

where bhp is the brake horsepower of the engine and $C_1 = C_2 = 0$ when no air cleaner is installed. When an air cleaner has been installed $C_2 = C_1 + 7$ dB and $C_1 = 13 - 5\delta_s + 8\delta_f$ dB where $\delta_s = 1$ if there is a snorkel on the air cleaner, otherwise $\delta_s = 0$; and $\delta_f = 1$ if there is a frontal intake on the air cleaner, otherwise $\delta_f = 0$.

Truck Tire Noise[31,32]

An estimate of the A-weighted-sound-pressure level at 15 m for truck tires can be obtained from the relation

$$L_T = 40\log_{10} V + 10\log_{10} N_a L_o - B_o \qquad \text{dB re 20 } \mu\text{Pa} \qquad (5\text{-}38)$$

where N_a is the number of axles, L_o is the load per tire (kg), V is the velocity of the vehicle (km/hr), and B_o is a function of the tread design and wear condition and is given in Table 5-17. The values in Table 5-17 are independent of road surface texture except graded asphaltic concrete, pitted Portland cement concrete, and any wet surface. Equation 5-38 is valid to within ±3 dB.

TABLE 5-17
Values for B_o in Equation 5-38

Tire tread	B_o (dB) New	Worn
Neutral rib	36.5	—
Rib	35.0	32.5
Cross lug	30.5	26.0
Pocket retread	19.5	19.5

Equation 5-38 can also be used to estimate the A-weighted levels at 15 m for mixed-matched sets of tires in the following manner:

(1) to estimate the noise due to only the outside tire of a matched set, subtract 1.5 dB from the loaded, four-tire, single-axle level;
(2) to estimate the noise due to the inside tire of a matched set, subtract 6 dB from the loaded, four-tire single-axle level;
(3) to estimate the noise attributable to opposite side tires, subtract 3 dB from the loaded, four-tire, single-axle level;
(4) to arrive at an estimate of the noise of a mixed set of tires, add the individual contributions (Steps 1, 2, and 3);
(5) for a second loaded axle, add 3 dB; and
(6) for separated axles, subtract 2 dB from the quieter set and add according to (1-24).

Truck Transmission Noise[31,32]

An estimate of the A-weighted sound-pressure level at 15 m from truck transmissions is obtained from the following relation:

$$L_G = 10 \log_{10} \text{bhp} + 13.5 \log_{10} N - 2.7 \qquad \text{dB re 20 } \mu\text{Pa} \qquad (5\text{-}39)$$

where bhp is the brake horsepower of the engine and N is the engine speed (rpm). Equation 5-39 is valid to within ±6 dB over the speed range of 1500–2100 rpm.

Example 5-17. A 20,000 kg truck has ten new cross-lug tires on three axles and is traveling at 80 km/hr. The diesel engine is four-stroke and naturally aspirated, with an air cleaner plus snorkel intake. The exhaust contains a standard muffler. At this speed the brake horsepower of the engine is 265 at 1900 rpm. The engine has a displacement of 6000 cm^3 and a bore of 15 cm. The fan consists of 8 blades, 0.4 m in diameter and 0.07 m wide. Estimate the total A-weighted-sound-pressure level at 15 m.

Solution: From (5-30) the A-weighted level from the engine is

$$L_E = 30\log_{10}1900 + 50\log_{10}15 - 70.7 = 86.5 \text{ dB re } 20 \text{ } \mu\text{Pa}$$

The A-weighted level for the muffled exhaust is estimated from (5-33), thus

$$L_M = 10\log_{10}265 + 74.5 - 17.2 = 81.5 \text{ dB re } 20 \text{ } \mu\text{Pa}$$

The A-weighted level for the fan is obtained from (5-34), thus

$$L_F = 10\log_{10}(0.07)(8) + 30\log_{10}\left\{[(1900)(0.4)]^2 + [(5.305)(80)]^2\right\}$$
$$- 108.6 = 65.3 \text{ dB re } 20 \text{ } \mu\text{Pa}$$

The A-weighted intake noise level is estimated from (5-36), hence

$$L_I = 81 - (13 - 5) = 73 \text{ dB re } 20 \text{ } \mu\text{Pa}$$

The A-weighted tire noise level is estimated from (5-38), thus

$$L_T = 40\log_{10}80 + 10\log_{10}(3)(2000) - 30.5 = 83.4 \text{ dB re } 20 \text{ } \mu\text{Pa}$$

wherein it has been assumed that the total load has been distributed equally to each tire. The A-weighted transmission noise level is estimated from (5-39), thus

$$L_G = 10\log_{10}265 + 13.5\log_{10}1900 - 2.7 = 65.8 \text{ dB re } 20 \text{ } \mu\text{Pa}$$

Using (1-24) the total estimated A-weighted-sound-pressure level at 15 m is found to be

$$L_{\text{TOTAL}} = 10\log_{10}(10^{8.65} + 10^{8.15} + 10^{6.53} + 10^{7.3} + 10^{8.34} + 10^{6.58})$$
$$= 89.2 \text{ dB re } 20 \text{ } \mu\text{Pa}$$

Train Noise[33-35]

From measurements on stationary engines at maximum throttle settings the following expression for the estimated A-weighted-sound-pressure level at 15 m from an unmuffled exhaust has been obtained:

$$L_A = 10\log_{10}\text{hp} + 63.2 - T_o, \qquad \text{dB re } 20 \text{ } \mu\text{Pa} \qquad (5\text{-}40)$$

where hp is the horsepower of the engine and $T_o = 6$ dB for turbocharged engines and $T_o = 0$ otherwise. It is mentioned that most diesel locomotives run with unmuffled exhausts.

The A-weighted-sound-pressure level due to engine casing noise at 15 m can be estimated from the following relation:

$$L_A = 30 \log_{10} N + 10 \log_{10} \text{hp} - 31.4 \qquad \text{dB re 20 } \mu\text{Pa} \qquad (5\text{-}41)$$

where N is the engine speed and hp is the engine horsepower.

The A-weighted-sound-pressure level at 30 m due to wheel-rail interaction on tangent, slightly curved track can be estimated from the relation

$$L_A = 26.8 + 30 \log_{10} V \qquad \text{dB re 20 } \mu\text{Pa} \qquad (5\text{-}42)$$

where V is the velocity of the cars (km/hr).

5.8 A NOMOGRAPH METHOD FOR THE PREDICTION OF HIGHWAY TRAFFIC NOISE

Introduction

The Federal Highway Administration of the U. S. Department of Transportation has developed three procedures whereby traffic noise from freely flowing highway traffic can be reasonably well-predicted.[36] Two of these procedures are graphical and the third requires a digital computer program. In this section we present the graphical procedure that uses a nomograph. This nomograph predicts the A-weighted L_{10} level and is valid for moderately high volume freely flowing traffic on infinitely long, unshielded, straight, level roadways. Adjustments are then made to the values obtained from the nomograph to include some of the effects of roadway geometry and road surface characteristics.

There are many situations where the traffic flow is intermittent, where cars and trucks operate in accelerating and decelerating modes, or where the principal sound source is an intermittent line of low speed, low volume trucks climbing a steep ramp grade. Simple and reliable noise prediction schemes for such complicated situations are not yet available.

Information Required to Use the Nomograph

To use the nomograph, shown in Figure 5-23, the following information is required: the total number of automobiles and trucks per hour, the percentage of this total volume flow that comprise trucks, the average speed of the vehicles, and the various distances from the centerline of the

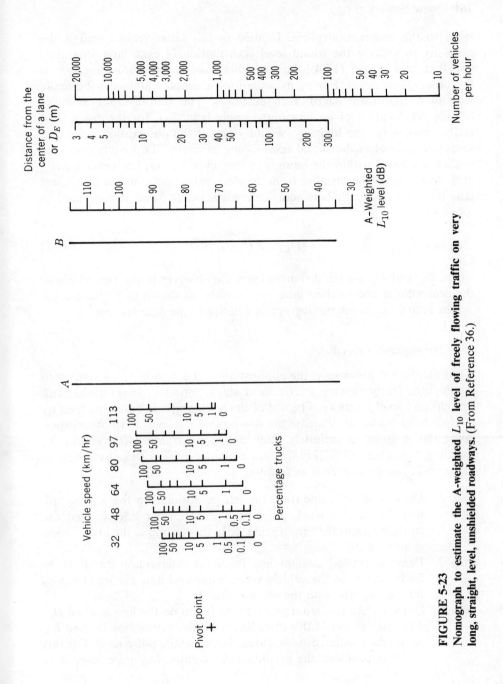

FIGURE 5-23
Nomograph to estimate the A-weighted L_{10} level of freely flowing traffic on very long, straight, level, unshielded roadways. (From Reference 36.)

163

lanes to the observation point. Implied in this latter requirement is the necessity to analyze the sound level contribution of each lane separately and then add [as per (1-24)] the individual lane contributions together at the observation point. For many situations a simplification can be made without a significant sacrifice in accuracy. The simplification involves finding the location of an imaginary single lane that, for the given total traffic volume for the highway, would yield, at the observation point, the same sound level as the actual several-lane geometry. This equivalent lane is always located within the bounds of the several lanes, but never exactly at the centerline. The distance from the observation point to the equivalent lane is called the single-lane-equivalent-distance, D_E, and is computed as follows:

$$D_E = \sqrt{D_N D_F} \qquad (5\text{-}43)$$

where D_N and D_F are the distances from the observer to the centerlines of the nearest lane and farthest lane, respectively, as shown in Figure 5-24.

The actual use of the nomograph is detailed in the next section.

Nomograph Procedure

As mentioned previously, the application of the nomograph is limited to continuous, freely flowing traffic on a single infinitely long, unshielded, straight and level roadway. The use of the nomograph is best explained by the following example: Estimate the A-weighted L_{10} level at an observation point 150 m from an infinitely long highway carrying 2400 vehicles/hr traveling at a speed of 97 km/hr and containing 5% trucks. Referring to Figure 5-25 the procedure is as follows:

Step 1. Draw a straight line from the left pivot point on the nomograph through the 5% truck point on the 97 km/hr line. Extend the straight line to the turning line A. In this example, the intersection is marked $A1$.

Step 2. Draw a second straight line from the intersection point $A1$ to 2400 veh/hr on the vehicle volume line and note the intersection, $B1$, of this line with the vertical line B.

Step 3. Draw a third line from point $B1$ to 150 m on the line marked D_E. The intersection of this third line with the vertical line labeled L_{10} yields the predicted A-weighted, 10-percentile noise level. For this example problem, the predicted A-weighted L_{10} noise level is 71 dB.

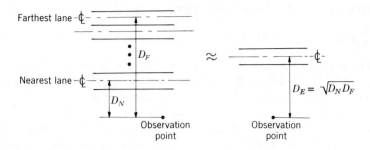

FIGURE 5-24
Pictorial representation of the single-lane-equivalent-distance, D_E.

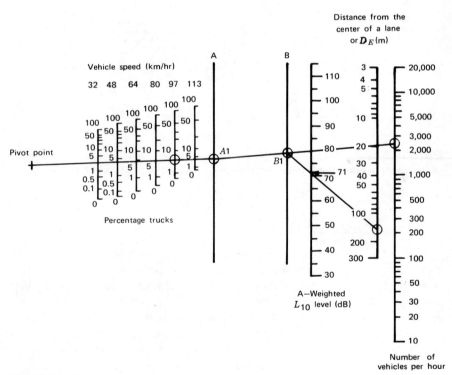

FIGURE 5-25
Illustration of the use of the nomograph of Figure 5-23.

Adjustments to the Nomograph Value

Road Segment

For practical purposes, a road segment can be considered an infinitely long highway if it extends in each direction a distance of at least four times the observation near-lane distance. If the segment does not meet this criterion, an adjustment is made to decrease the L_{10} level because the segment is finite. The amount of this decrease is obtained from Figure 5-26. This figure is also valid for semi-infinite roadway segments by simply letting the line labeled A in the inset to Figure 5-26 be parallel to the roadway before determining the angle θ.

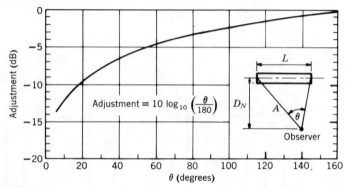

FIGURE 5-26
Adjustment to nomograph value for roadways of finite length. (From Reference 36.)

Road Surface and Grade

For vehicles traveling on very rough or very smooth pavement, the basic noise level computations are adjusted upward or downward, as the case may be, by 5 dB, in accordance with Table 5-18. Only rarely should such an adjustment be applied to truck noise, and then only upward for trucks traveling at speeds above 100 km/hr and when the pavement is particularly rough. For the great majority of new surfaces no adjustment is warranted. Occasionally an old surface, worn badly by studded tires or intentionally grooved, is encountered for which a 5 dB positive adjustment is justified. Less frequently, a very smooth-coated surface warrants a 5 dB negative adjustment. Such smooth surface roads, however, are rare because of their inherent low friction characteristics.

TABLE 5-18
Adjustments to Automobile Noise Levels for Various Road Surfaces

Type of surface	Description	Adjustment (dB)
Smooth	Very smooth, seal-coated asphalt pavement	−5
Normal	Moderately rough asphalt and concrete surface	0
Rough	Rough asphalt pavement with large voids (13 mm or larger in diameter) and grooved concrete	+5

The positive adjustments to account for the increased noise of trucks on gradients are shown in Table 5-19. These adjustments are made only to truck noise levels, and are never negative, that is, there is no adjustment for a downhill gradient. In most situations where the two-directional lanes appear together on a gradient, the adjustment may be applied equally to both sides of the highway without regard to whether the near lane is an up-gradient or a down-gradient.

TABLE 5-19
Adjustments to Truck Noise Levels
for Various Road Gradients

Gradient (%)	Adjustment (dB)
⩽2	0[a]
3–4	+2
5–6	+3
⩾7	+5

The adjustments apply to either the trucks alone or the automobiles alone. Hence the nomograph must be used other than indicated on p. 164 when any of these adjustments is different from zero. The difference lies in the fact that now the L_{10} level must be obtained first for the automobiles alone and then for the trucks alone. For the automobiles alone the percentage of trucks is zero and the volume flow used in the nomograph corresponds to the number of automobiles per hour. Similarly for the trucks alone the percentage of trucks is 100 and the volume flow used in the nomograph corresponds to the number of trucks per hour.

These ideas are summarized in the tally sheet shown in Table 5-20.

TABLE 5-20
Traffic Noise Tally Sheet

Observer distance or segment						
Vehicle type	Auto	Truck	Auto	Truck	Auto	Truck
A-weighted L_{10} level (Figure 5-23) (dB)						
Segment (Figure 5-26) (dB)						
Gradient (Table 5-19) (dB)						
Road surface (Table 5-18) (dB)						
L_{10} at observer distance by vehicle type (dB)						
L_{10} at observer (Figure 1-9) (dB)						
L_{10} segment total, if applicable (Equation 1-24) (dB)						

Example 5-18. Consider a straight highway that has three lanes in each direction separated by a 7.5 m wide median strip. The highway has a 5% gradient and the pavement is rough asphalt. The highway carries 6700 autos/hr and 500 trucks/hr at a speed of 97 km/hr. There are two locations of interest: #1 is 30 m and #2 is 150 m from the center line of the nearest lane. Each lane is 3.8 m wide. Obtain an estimate of the A-weighted L_{10} level.

Solution: The roadway width is 30.3 m. The equivalent lane distance for the 30 m and 150 m observation points are, respectively, $D_E = \sqrt{(56.5)(30)}$ $= 41.2$ m and $D_E = \sqrt{(176.5)(150)} = 162.7$ m. Using the tally sheet given in

Table 5-20 as a guide, the A-weighted L_{10} levels at the two observation points are determined as shown in Table 5-21.

TABLE 5-21
Tally Sheet for Example 5-18

Observer distance from nearest lane (m)	30		150	
Vehicle type	Auto	Truck	Auto	Truck
A-weighted L_{10} level (Figure 5-23) (dB)	75	79	71	75
Gradient (Table 5-19) (dB)	0	+ 3	0	+ 3
Road surface (Table 5-18) (dB)	+ 5	0	+ 5	0
L_{10} at observer distance by vehicle type (dB)	80	82	76	78
L_{10} at observer (Equation 1-24) (dB)	84		80	

Example 5-19. Consider a highway with the geometry shown in Figure 5-27 that has the same lane configuration, pavement, gradient, and traffic flow as Example 5-18. Estimate the A-weighted L_{10} level at the observer position shown in Figure 5-27. The distances shown are to the center line of the nearest lane.

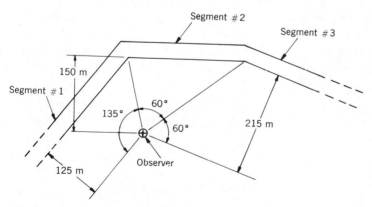

FIGURE 5-27
Highway geometry for Example 5-19.

Solution: The equivalent lane distances are: for segment #1, $D_E = 137.6$ m; for segment #2, $D_E = 162.7$ m; and for segment #3, $D_E = 227.9$ m. The solution is tabulated in Table 5-22.

TABLE 5-22
Tabulation of the Solution to Example 5-19

Segment	#1		#2		#3	
Vehicle type	Auto	Truck	Auto	Truck	Auto	Truck
A-weighted L_{10} level (Figure 5-23)　(dB)	72	77	71	76	69	74
Segment (Figure 5-26)　(dB)	−1.3	−1.3	−4.8	−4.8	−4.8	−4.8
Gradient (Table 5-19)　(dB)	0	+3	0	+3	0	+3
Road surface (Table 5-18)　(dB)	+5	0	+5	0	+5	0
L_{10} at observer distance by vehicle type　(dB)	75.7	78.7	71.2	74.2	69.2	72.2
L_{10} at observer (Figure 1-9)　(dB)	80.5		76		74	
L_{10} segment total (Equation 1-24)　(dB)	82.5					

5.9　COOLING TOWER NOISE

A method has been devised[37] whereby the A-weighted and octave band sound-pressure levels can be accurately predicted both at the base and at any distance from a natural-draft cooling tower. A typical cross-section of each tower is shown in Figure 5-28. The A-weighted sound level at the rim of the pond can be obtained from the expression

$$p_{\text{rim}} = \sqrt{W_{\text{ac}} \frac{\rho c}{2\pi R h'}} \qquad \text{Pa} \qquad (5\text{-}44)$$

where

$$W_{\text{ac}} = Mh \left[0.95 \left(\frac{T}{h} \right)^2 + 1.8 \left(\frac{D}{h} \right)^2 \right] \times 10^{-5} \qquad \text{W} \qquad (5\text{-}45)$$

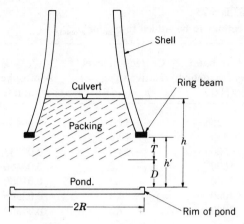

FIGURE 5-28
Typical natural-draft cooling tower construction. (Reprinted, by permission, from Reference 37.)

and M is the mass flow rate of the cooling water (kg/sec), h is the distance the water falls from the culvert to the pond (m), T is the depth of the packing below the ring beam (m), D is the height from the pond to the base of the packing (m), h' is the depth of the open area below the tower shell (m), R is the radius of the pond (m), and ρc is the characteristic impedance of the air ($=412$ N-sec/m^3 at standard conditions).

The A-weighted sound level at a distance A (m) from the rim of the pond is

$$p_{\text{pond}} = \left[W_{\text{ac}} \frac{\rho c}{\pi A^2 \left(1 + \dfrac{2R}{A}\right)} \tan^{-1} \sqrt{1 + \frac{2R}{A}} \right]^{1/2} \qquad \text{Pa} \qquad (5\text{-}46)$$

To determine the octave band pressure levels at either the rim of the pond or at the distance A from the rim the following values given in Table 5-23 are added to the A-weighted levels determined by either (5-44) or (5-46), as the case may be. For distances greater than 30 m from the rim ($A > 30$ m) the octave band levels should be corrected for atmospheric absorption as shown in the third column of Table 5-23.

TABLE 5-23

Corrections to be Added to p_{rim} or p_{pond}

Octave band center frequency (Hz)	Correction level relative to p_{rim} or p_{pond} (dB)	Atmospheric absorption (dB/m)
125	−19.4	—
250	−19.8	—
500	−13.0	0.00233
1000	−7.8	0.00466
2000	−6.3	0.01000
4000	−4.3	0.02566
8000	−7.2	0.04800

Example 5-20. A natural-draft tower has a water flow rate of 8500 kg/sec from a culvert 11.6 m above the level of the pond. The packing is 8.22 m thick and it extends down to the pond. The open height is 7.92 m and the radius of the pond is 47.2 m. In addition there is warm water at the rate of 1510 kg/sec being sprayed from a pipe that runs around the perimeter of the tower 7.32 m above the pond. Estimate the A-weighted sound level and the octave band levels at the rim of the pond and 60 m from the rim.

Solution: Using (5-45) the tower water creates an acoustic level of

$$W_{ac}^{(T)} = (8500)(11.6)(0.95)\left(\frac{8.22}{11.6}\right)^2 \times 10^{-5} = 0.470 \text{ W}$$

wherein we have used the fact that $D=0$. The warm water flow creates an acoustic level of

$$W_{ac}^{(P)} = (1510)(7.32)(1.8 \times 10^{-5}) = 0.199 \text{ W}$$

since in this case $T=0$ and $h=D=7.32$ m. The total acoustic power is $W_{ac} = W_{ac}^{(T)} + W_{ac}^{(P)} = 0.669$ W.

The A-weighted sound level at the rim is, from (5-44) and (1-18),

$$L_A = 10\log_{10}\frac{(0.669)(412)}{(2\pi)(47.2)(7.92)} + 94 = 84.7 \text{ dB re } 20 \text{ } \mu\text{Pa}$$

The A-weighted sound level at 60 m from the rim is, from (5-46) and (1-18),

$$L_A = 10\log_{10}\frac{(0.669)(412)}{\pi^2(3600+5664)}\tan^{-1}\sqrt{1+\frac{94.4}{60}} + 94$$

$$= 68.8 \text{ dB re } 20 \text{ } \mu\text{Pa}$$

The octave band levels at the rim of the pond are obtained using Table 5-23. For example, at 250 Hz we have $84.7 - 19.4 = 65.3$ dB. The octave band level at a distance of 60 m from the rim of the pond is also obtained from Table 5-23 except that the atmosphere corrections are also used. For example, at 500 Hz we have $68.8 - 13.0 - (60)(0.00233) = 55.7$ dB. The results are summarized in Table 5-24.

TABLE 5-24
Tabulation of the Solution to Example 5-20

Octave band center frequency (Hz)	Octave band levels	
	At rim (dB)	60 m from rim (dB)
125	65.3	49.4
250	64.9	49.0
500	71.7	55.7
1000	76.9	60.7
2000	78.4	61.9
4000	80.4	63.0
8000	77.5	58.7

REFERENCES

1. ASHRAE Handbook and Product Directory—Systems, American Society of Heating, Refrigeration, and Air Conditioning Engineers, New York (1973), Chapter 35, pp. 503–511.

2. G. C. Groff, J. R. Schreiner, and C. E. Bullock, "Centrifugal Fan Sound Power Level Prediction," *ASHRAE Trans.*, Vol. 73, Part II (1967), pp. V.4.1–V.4.18.

3. J. B. Graham, "How to Estimate Fan Noise," *Sound Vib.* (May 1972), pp. 24–27.

4. R. M. Hoover and C. O. Wood, "Noise Control for Induced Draft Fans," *Sound Vib.* (April 1970), pp. 20–24.

5. C. M. Harris, Ed., *Handbook of Noise Control*, McGraw-Hill, New York (1957), pp. 25–10 to 25–11.

6. H. C. Simpson, T. A. Clark, and G. A. Weir, "A Theoretical Investigation of Hydraulic Noise in Pumps," *J. Sound Vib.*, Vol. 5, No. 3 (1967), pp. 456–488.

7. "Handbook for Shipboard Airborne Noise Control," Submitted by Bolt, Beranek and Newman under Contract No. DOT-CG-20756A to U. S. Coast Guard Headquarters, Washington, D. C. and Naval Ship Engineering Center, Hyattsville, Md. (February 1974).

8. "Noise Control for Mechanical Equipment," Department of the Army Technical Manual TM 5-805-4, Headquarters, Department of the Army, Washington, D. C. (September 1970).

9. I. Heitner, "How to Estimate Plant Noises," *Hydrocarbon Proc.*, Vol. 47, No. 12 (December 1968), pp. 67–74.

10. W. E. Blazier, Jr., "Noise from Large Centrifugal Compressors," in *Noise and Vib. Control Engineering*, M. J. Crocker, Ed., Proceedings of the Purdue Noise Control Conference, Purdue University, Lafayette, Indiana (1972), pp. 84–89.

11. C. E. Bullock, "Aerodynamic Sound Generation by Duct Elements," *ASHRAE Trans.*, Vol. 76, Part II (1970), pp. 97–108.

12. I. L. Vér, "Prediction Scheme for the Self-Generated Noise of Silencers," *Inter-Noise 72 Proc.*, International Conference on Noise Control Engineering, Washington, D. C. (1972), pp. 294–298.

13. L. L. Beranek, Ed., *Noise and Vibration Control*, McGraw-Hill, New York (1971), pp. 522–528.

14. G. Reethof and A. V. Karvelis, "Control Valve Noise and Its Reduction—State of the Art," Ref. 12, pp. 146–153.

15. F. J. Heymann, "Some Experiments Concerning Control Valve Noise," Inter-Noise 73 Proceedings (October 1973), pp. 382–388.

16. A. V. Karvelis and G. Reethof, "Valve Noise Research Using Internal Wall Pressure Fluctuations," Inter-Noise 74 Proceedings (September/October 1974), pp. 331–336.

17. A. Nakano, "Characteristics of Noise Emitted by Valves," Paper F-5-7, 6th International Congress on Acoustics, Tokyo, Japan (1968).

18. C. B. Schuder, "Coping with Control Valve Noise," *Chem. Eng.*, (October 19, 1970).

19. C. B. Schuder, "Control Valve Noise—Prediction and Abatement," Ref. 10, pp. 90–94.

20. E. E. Allen, "Mechanics of Noise Generation by Fluid Flow Through Central Valves," Ref. 12, pp. 299–304.

21. E. E. Allen, "Prediction and Abatement of Control Valve Noise," Paper No. 69-535, Proceedings of the Annual Conference of the Instrument Society of America (1969).

22. J. W. Hutchison, Ed., *ISA Handbook of Control Valves*, Instrument Society of America, Pittsburg, Pa. (1971), pp. 136–167.

23. H. D. Baumann, "On the Prediction of Aerodynamically Created Sound Pressure Level of Control Valves," ASME Paper No. WA/FE-28 (December 1970).

24. "Masoneilan Low Noise Control Equipment," Bulletin No. 340E (1971), Masoneilan International Inc., Norwood, Mass.

25. A. J. King, *The Measurement and Supression of Noise*, Chapman and Hall, London (1965), pp. 81–87, 102.

26. "Power Plant Acoustics," Department of the Army Technical Manual TM 5-805-9, Headquarters, Department of the Army, Washington, D. C. (December 1968).

27. L. D. Mitchell, "Gear Noise: The Purchaser's and Manufacturer's View," Ref. 10, pp. 95–106.

28. A. Y. Attia, "Noise of Gears of Circular-Arc Tooth Profile," *J. Sound Vib.*, Vol. 11, No. 4 (1970), pp. 383–397.

29. "Noise from Construction Equipment and Operation, Building Equipment and Home Appliances," Report No. NTID 300.1, U. S. Environmental Protection Agency, Washington, D. C. (December 31, 1971).

30. "A Study of Noise-Induced-Hearing-Damage Risk for Operators of Farm and Construction Equipment," Technical Report, SAE Research Project R-4, Society of Automotive Engineers, Inc., New York (December 1969).

31. P. E. Rentz and L. D. Pope, "Description and Control of Motor Vehicle Noise Sources," Vol. 2 of "Establishment of Standards for Highway Noise Levels," Report No. NCHRP 3-7/3, Highway Research Board, National Cooperative Highway Research Program, National Academy of Sciences, Washington, D. C. (February 1974) (Draft Report).

32. P. E. Waters, "Commercial Road Vehicle Noise," *J. Sound Vib.*, Vol. 35, No. 2 (1974), pp. 155–222.

33. E. K. Bender, R. A. Ely, P. J. Remington, and M. J. Rudd, "Railroad Environmental Noise: A State-of-the-Art Assessment," Bolt, Beranek and Newman Report No. 2709 to the association of American Railroads, Washington, D. C. (January 1974).

34. J. E. Manning, R. G. Cann, and J. J. Fredberg, "Prediction and Control of Rail Transit Noise and Vibration—A State-of-the-Art Assessment," Cambridge Collaborative Inc. Interim Report to Department of Transportation, Urban Mass Transportation Administration, Washington, D. C. (March 1974).

35. J. W. Swing and D. B. Pies, "Assessment of Noise Environments Around Railroad Operations," Wyle Laboratories Report WCR 73-5 for Southern Pacific Transportation Co., et al. under Contract No. 0300-94-07991 (July 1973).

36. "Fundamentals and Abatement of Highway Traffic Noise," Federal Highway Administration Report No. FHWA-HH1-HEV-73-7976-1 (June 1973), Chapter 4.

37. R. M. Ellis, "Cooling Tower Noise Generation and Radiation," *J. Sound Vib.*, Vol. 14, No. 2 (1971), pp. 171–182.

6

ROOM ACOUSTICS

6.1 SOUND ABSORPTION COEFFICIENTS

Introduction

When sound waves fall on a surface or object, their energy is partially reflected and partially absorbed. The sound-absorbing efficiency of the surface involved is given in terms of an absorption coefficient designated by the symbol α. There are several types of absorption coefficients and they are described in this section.

Sound-Absorption Coefficient for a Given Angle of Incidence, α_θ.

The sound-absorption coefficient α_θ is defined as the ratio of the sound energy absorbed by a surface to the sound energy incident upon that surface at a given angle θ. The most common type of angle of incidence is normal incidence, $\theta = 0$. This coefficient, α_0, sometimes denoted α_n, is measured[1] in a standing-wave tube (sometimes called an impedance tube) shown in Figure 6-1. An audio signal generator drives a loudspeaker that transmits plane waves longitudinally along the tube. Waves of reduced amplitude are reflected by the specimen and combine with the incident waves to form a standing wave pattern along the tube. This pattern is explored by the moveable microphone or probe on the axis of the tube, whose output is fed through an electrical filter to an output indicating device that shows the relative pressure amplitudes at the maximums and minimums in the standing wave pattern, denoted p_{max} and p_{min}, respectively. An additional requirement for determining α_{ST} (see p. 178) is a

176

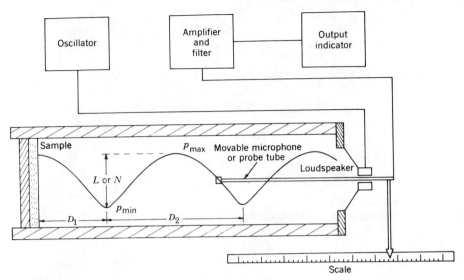

FIGURE 6-1
Schematic diagram of apparatus and equipment necessary to make a measurement in
a standing wave tube.

calibrated scale indicating the position of the microphone or probe in the
standing wave pattern with respect to the face of the specimen.

The normal incident coefficient, α_n, is determined from the relationship

$$\alpha_n = \frac{4N}{(N+1)^2} \tag{6-1}$$

where

$$N = \left| \frac{p_{max}}{p_{min}} \right| = 10^{(L/20)} \tag{6-2}$$

and L is the difference in dB between the maximum and minimum
sound-pressure levels in the standing wave pattern in the tube; that is,
$L = 20\log_{10}N$. From (6-1) and (6-2) it is seen that when $L \geqslant 32$ dB,
$p_{max} \geqslant 40 p_{min}$, $\alpha_n < 0.1$, and the material is, for all practical purposes, a
perfect reflector of sound. Equation 6-1 is plotted in Figure 6-2.

The measurement of α_n in the impedance tube by means of a tone-burst
technique has recently been developed[2] to work over the frequency range
of 500–10,000 Hz at peak sound-pressure levels of 165 dB re 20 μPa. These
peak levels are not attainable with the standing wave tube described above.

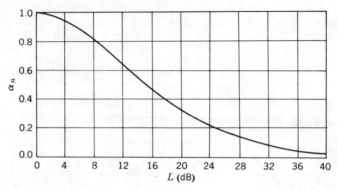

FIGURE 6-2
Normal incidence absorption coefficient as a function of L.

Statistical (Energy) Sound-Absorption Coefficient, α_{ST}.

With a view to obtaining a single number index for general use, α_{ST} is defined (for an absorbing surface of infinite extent) as the ratio of sound energy absorbed by the surface to the sound energy incident upon the surface when the incident sound field is diffuse. As mentioned earlier, a field is diffuse if a great many reflected waves cross a given point in space from all possible directions such that the sound energy density is uniform through the field around the point. The statistical sound-absorption coefficient is sometimes called the random-incidence sound-absorption coefficient.

The statistical absorption coefficient α_{ST} can be determined from the standing wave tube data in the following manner:[3,4]

$$\alpha_{ST} = \frac{8N(1+t^2)}{N^2+t^2}\left\{1 - \frac{N(1+t^2)}{N^2+t^2}\,ln_e\left[\frac{(N+1)^2(1+t^2)}{1+N^2+t^2}\right]\right.$$

$$\left. + \frac{N^2(1+t^2)^2 - t^2(1-N^2)^2}{t(N^2+t^2)(1-N^2)}\tan^{-1}\left[\frac{t(1-N)}{1+Nt^2}\right]\right\} \qquad (6\text{-}3)$$

where N is given by (6-2),

$$t = \cot\frac{2\pi D_1}{\lambda}$$

and $\lambda = 2D_2$ is the wavelength of the sound. Equation 6-3 is shown in Figure 6-3 as a function of D_1, λ, and L. It should be realized that the determination of α_{ST} requires the additional information D_1, which is not

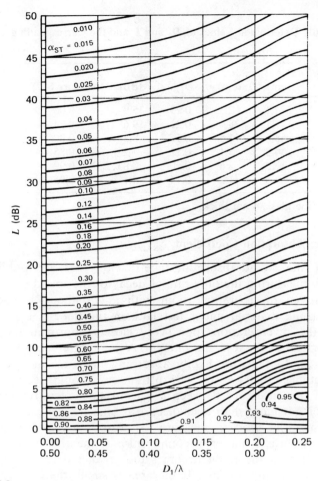

FIGURE 6-3
Statistical absorption coefficient as a function of D_1, λ. and L. (From Reference 2.)

required to determine α_n. From Figure 6-1 it is seen that D_1 is the distance from the face of the specimen to the first pressure minimum and D_2 is the distance between any two adjacent pressure minima (or maxima).

Example 6-1. Given the experimentally determined values of D_1 and L tabulated in Table 6-1 determine both the normal incidence absorption coefficient α_n and the statistical energy absorption coefficient α_{ST}.

Solution: The results are easily determined from Figures 6-2 and 6-3 and are presented in tabular form in Table 6-1.

TABLE 6-1
Experimentally Determined Values of D_1 and L and the Corresponding Values of α_n and α_{ST}^a

Frequency (Hz)	λ (cm)	D_1 (cm)	D_1/λ	L (dB)	α_n (Figure 6–2)	α_{ST} (Figure 6–3)
125	274.1	62.5	0.23	35.9	0.06	0.11
250	137.1	28.4	0.21	19.6	0.34	0.44
500	68.6	11.5	0.17	8.3	0.80	0.78
1000	34.3	4.1	0.12	7.2	0.85	0.75
2000	17.2	8.6	0.50	0.83	0.99	0.89
4000	8.6	1.2	0.14	4.1	0.95	0.88

^{a}Material is fiberglass, 4 cm thick.

Sabine Absorption Coefficient, α_{SAB}

Most of the sound-absorption coefficients that are published are obtained by measuring the time rate of decay of the sound-energy density in an approved reverberation room with and without a patch of the sound-absorbing material under test laid on the floor (see p. 208). The design of the reverberation room very closely approximates a diffuse reverberant field as the test environment. A great difference exists between α_{ST} and α_{SAB}. In fact, for highly absorbing materials α_{SAB} can exceed unity, sometimes by 20 or 30%. The diffraction of sound seems to account for most of the difference between α_{SAB} and α_{ST} at low frequencies, but probably for not more than a small part at high frequencies. The area of absorbing material used to measure its absorption coefficient in a reverberation room is a matter of compromise. If the area is very large, the total absorption in the room becomes too great and the incident sound field is not sufficiently diffuse. If the area is small, the correction of the measured values to account for diffraction becomes large.

It should be noted that α_θ, α_{ST}, and α_{SAB} are functions of frequency. They are usually determined at the geometric mean frequencies of the 1/3- or 1/1-octave bands from 125 to 4000 Hz. Furthermore, it has been shown[5] that the sabine absorption coefficients, α_{SAB}, measured in the reverberation room are generally consistent with the values calculated from the standing wave tube, α_n.

Noise Reduction Coefficient (NRC)

The NRC is the average (to the nearest multiple of 0.05) of α_{SAB} at 250, 500, 1000, and 2000 Hz.

Average Sabine Absorption Coefficient for a Room, $\bar{\alpha}_{SAB}$

The most widely used room-average absorption coefficient is calculated by weighting the Sabine absorption coefficients $(\alpha_{SAB})_i$ of the individual surfaces of the room, S_i, according to the following formula:

$$\bar{\alpha}_{SAB} = \frac{1}{S} \sum_{i=1}^{N} S_i (\alpha_{SAB})_i \qquad (6\text{-}4)$$

where $S = \sum_{i=1}^{N} S_i$ is the total area of all the surfaces.

Absorbing objects such as chairs, seats, tables, desks, and people, must be included when calculating $\bar{\alpha}_{SAB}$, but such objects have an ill-defined area. Hence it is common practice to assign a value of Sabine absorption A_j to each object, where the absorption is equivalent to $S_j(\alpha_{SAB})_j$. All values of absorption for people and objects are summed, $A_0 = \sum A_j$, and are added to the numerator of (6-4). No modification to the total area is made.

6.2   POROUS MATERIALS

Introduction

A porous material is characterized by a number of variables. The first is porosity, which is the fraction of empty space within the material to its total volume. The second is the flow resistance, which is a measure of the difficulty with which air can be blown through a unit thickness of the material. The third is the structure factor, which is a measure of the amount of dead space, such as pores running parallel to the surface compared with pores that convey air through the material. A material with a structure factor of 1 has all its pores running parallel with the direction of incident sound. Materials with more tortuous pores have structure factors greater than 1. These ideas are illustrated in Figures 6-4 to 6-6.

Specific Flow Resistance and Flow Resistivity

The specific (unit area) flow resistance of any layer of porous material is defined as

$$R_f = \frac{\Delta p}{u} \qquad \text{N-sec/m}^3 \qquad (6\text{-}5)$$

where Δp is the applied air-pressure differential measured between the two sides of the layer in N/m^2, and u is the particle velocity through and perpendicular to the two faces of the layer in m/sec.

Low porosity   High porosity

FIGURE 6-4
Variation of porosity with flow resistance and structure factor held constant.

Low resistance   High resistance

FIGURE 6-5
Variation of flow resistance with porosity and structure factor held constant.

Low structure factor   High structure factor

FIGURE 6-6
Variation of structure factor with porosity and flow resistance held constant.

For bulk materials the flow resistivity (specific flow resistance per unit thickness of material) is

$$R_1 = \frac{R_f}{d} \qquad \text{N-sec/m}^4 \tag{6-6}$$

where d is the thickness of the material.

For materials with a high porosity (near unity) the flow resistivity, R_1, has been experimentally related[6] to the normal incidence absorption coefficient α_n by the relation

$$\alpha_n = \frac{4R}{(R+1)^2 + X^2} \tag{6-7}$$

where R and X are given by

$$R = 1 + 0.0571 \left(\frac{\rho f}{R_1} \right)^{-0.754}$$

$$X = -0.0870 \left(\frac{\rho f}{R_1} \right)^{-0.732}$$

and ρ is the density of air in kg/m^3 and f is the frequency. Equation 6-7 is plotted in Figure 6-7 as a function of the dimensionless parameter ($\rho f / R_1$).

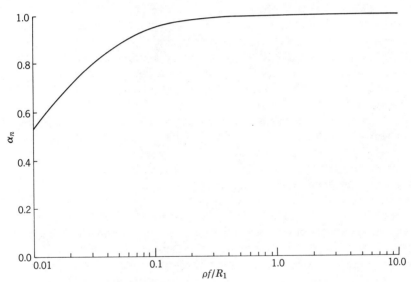

FIGURE 6-7
Normal incidence absorption coefficient given by (6-7) as a function of $\rho f / R_1$.

6.3   SOUND ABSORBING MATERIALS*

Introduction

Many materials absorb by reason of their porosity, whereby sound entering the interconnecting pores is dissipated into heat by the action of

*The tabulation of the sabine absorption coefficients for numerous commercially available products under a variety of mounting configurations are available from references 7–9. Reference 7 supplied much of the information in this section, which is used with the permission of the Controller of Her Britannic Majesty's Stationary Office.

viscous and thermal processes. Some porous materials are of the solid type, such as acoustic plaster or wood-wool board. Others are fibrous, such as acoustic felt, mineral wool or sprayed asbestos and the vibration of the fibrous material may contribute to the acoustic absorption. Porous absorbents are more efficient at high than at low frequencies. Cellular materials in which the air cells are enclosed and separate are not as effective as porous absorbents.

As distinct from the porous absorbent, there is the nonporous panel type in which vibration of the panel is the principal mechanism of absorption. Wood panelling is an example, and absorbents of this type have been used with thin membranes or flexible sheets as the vibrating surface. The panel absorber is mainly effective at low frequencies, and the absorption is of the resonant type. The resonance frequency can be controlled by choice of the panel weight and the depth of the air space behind the panel, and the absorption can be augmented by the incorporation of porous absorbent in the space behind the nonporous cover. Many treatments combine both types of absorption, as in the case of porous board mounted on battens.

A treatment that has attained great importance is the perforated panel absorbent, which functions as an array of Helmholtz resonators. It consists of a perforated panel backed by porous material, which may be placed in the space behind the panel, or alternatively take the form of a thin porous sheet stuck to the back of the panel. The Helmholtz resonator panel absorbent provides a flexible means of controlling the frequency characteristics of the absorption. The resonance frequency can be varied by the choice of panel thickness, the size and spacing of the holes, and the depth of the space behind the panel. The amount of the absorption and its frequency range can be controlled by the absorbing material employed.

Porous Absorbents

The general characteristics of porous felt-like materials are illustrated in Figure 6-8,* where the curves show the typical increase of absorption with frequency (recall Figure 6-7). When such materials are employed for acoustic correction it is important to use an adequate thickness if absorption over a wide frequency range is required. A thin porous absorbent applied directly to a wall surface works mainly at high frequencies and is inefficient at low frequencies. Figure 6-8 shows that the absorption is maintained further into the low frequency region when the thickness is increased. Mounting the material on battens to give an air space behind it acts in the same manner as an increase of thickness.

*Figures 6-8 through 6-19 are used with the permission of the Controller of Her Britannic Majesty's Stationery Office.

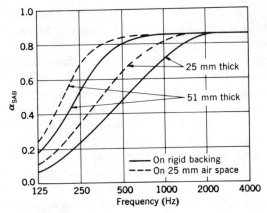

FIGURE 6-8
Absorption coefficient for 25 and 51 mm thick mineral wool slabs, each on a rigid backing and on 25 mm battens. (Reprinted, by permission, from Reference 7.)

Where the porous material is solid and has an open structure, maxima and minima of absorption can occur, as shown in Figure 6-9 for a wood-wool board. In this case the attenuation and the velocity of sound in the material are low, and the maxima and minima are associated with the interference between the sound reflected from the front surface and that which emerges from the material after reflection at the backing surface. The maxima and minima are displaced in the low frequency direction when the material is mounted on battens, and an increase of thickness

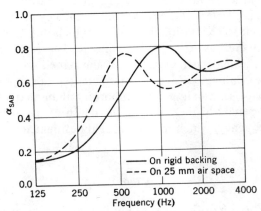

FIGURE 6-9
Absorption coefficient for wood-wool slabs, 38 mm thick, on a rigid backing, and on 25 mm battens. (Reprinted, by permission, from Reference 7.)

would produce a similar effect. With this type of material it is useful to employ a combination of different thicknesses and mounting conditions so as to average out the fluctuations in absorbing power.

The properties of porous absorbents are dependent on the surface treatment. Acoustic plasters require care in decoration and lose their efficiency if the surface is sealed by painting. Sprayed asbestos can be given various surface finishes, which should be in accordance with the specification of the manufacturer. Mineral wool is often covered with thin muslin. Acoustic felts can be covered with muslin that is painted and later pin-pricked. Porous materials are frequently covered with thin perforated sheet metal. With small perforations of 3 mm or less in diameter, and an open area of about 15% of the total surface area, the absorption of the porous material is not appreciably reduced by the cover (see next section). A nonporous membrane can be used as a cover, provided it is flexible and light enough. The principal effect of such a membrane is to reduce the absorption at high frequencies. The frequency at which the reduction becomes appreciable can be estimated roughly by the relation

$$f = \frac{98}{m} \quad \text{Hz} \tag{6-8}$$

where m is the weight per unit area of the membrane in kg/m^2.

Unperforated Panel Absorbents

The behavior of absorbents of this type may be understood from Figures 6-10 and 6-11, which give results for plywood panels mounted on battens. The absorption is due mainly to the flexural vibration of the plywood, and is greatest at the resonance frequency of the panel. The most important mode of vibration is that in which the panel vibrates as a diaphragm on the cushion of air behind it. The resonance frequency is governed by the mass per unit area of the panel, the stiffness of the air space and the bending stiffness of the panel on its supports. The stiffness of the air space is usually more important than that of the panel. An increase of the mass of the panel or the depth of the air space lowers the resonance frequency, which may be roughly estimated by the formula

$$f = \frac{1893}{\sqrt{dm}} \quad \text{Hz} \tag{6-9}$$

where d is the depth of the air space in mm and m the weight per unit area in kg/m^2. The above formula is for a flexible panel or membrane. The frequency might be higher by a factor of 50% for a relatively inflexible panel on supports close together.

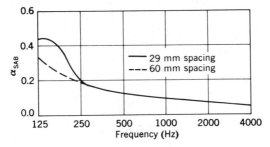

FIGURE 6-10
Absorption coefficient for 13 mm thick plywood panels spaced 29 and 60 mm from wall with absorbent in the space. (Reprinted, by permission, from Reference 7.)

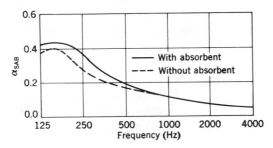

FIGURE 6-11
Absorption coefficient for 4.8 mm thick plywood panels spaced 51 mm from the wall, with and without absorbent in the space. (Reprinted, by permission, from Reference 7.)

Panel absorbents are usually designed for the low-frequency region and the absorption is in a relatively narrow range about the resonance frequency. The absorption and the bandwidth are increased by the use of porous absorbent behind the panel. Uniform absorption over a wider frequency range may be obtained by a combination of panels with different resonance frequencies. In addition to boardlike materials, such as plywood, flexible sheets and membranes have been used as covers for panel resonator absorbers. In buildings, constructions such as suspended ceilings, wooden floors, plaster on lath, and windows act as vibrating panels and contribute to the low-frequency absorption.

Porous materials mounted on battens frequently act both as porous and vibrating panel absorbers, and the absorption due to vibration helps to keep up the absorption at low frequencies. Figure 6-12 shows the absorption of insulation fiberboard under different mounting conditions. The porous absorption at high frequencies is combined with absorption due to vibration at low frequencies when the board is mounted on battens, to furnish a useful degree of absorption over a wide frequency range.

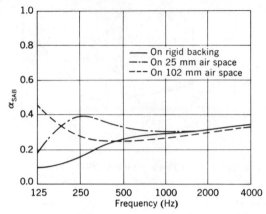

FIGURE 6-12
Absorption coefficient for 13 mm thick insulation fiberboard on a rigid backing and on 25 and 102 mm battens. (Reprinted, by permission, from Reference 7.)

Perforated Tiles and Boards

The typical features of the materials in this section are shown by the perforated tiles that are widely used as general purpose absorbents. The tiles are made of porous material, usually fiberboard or asbestos, and they may have a hard or nonporous face. The front surface is perforated with holes that penetrate into the porous interior. The sound absorbing properties are due mainly to the holes, and where these are effective the tiles can be painted without undue loss of efficiency, provided the holes are kept clear of paint.

Some of the properties of the tiles are illustrated in Figure 6-13, which gives average results for tiles of similar composition, but of different hole depth and thickness. The perforations are about 4.8 mm in diameter and about 13 mm apart. The holes act as resonators, giving an absorption peak; the resonance frequency increases as the hole depth decreases. Absorbents of this type tend to fall off in efficiency below about 500 Hz.

The absorption characteristics vary over a wide range according to the kind of porous material employed. Figure 6-14 gives results for three tiles of approximately the same geometry but of different materials. The tiles are mounted on battens in this case, giving rise to absorption due to panel vibration in the region of 250 Hz. The properties of tile A are similar to those discussed above. Tile C is of a denser, less porous material, with the result that the hole resonance frequency is high, and above the frequency range of the measurements. Tile B shows an intermediate type of characteristic in which the resonance absorption is less pronounced.

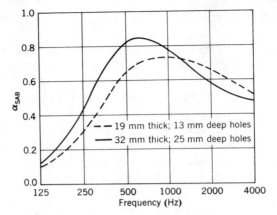

FIGURE 6-13
Absorption coefficient of two different perforated fiberboard tiles. (Reprinted, by permission, from Reference 7.)

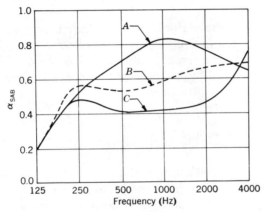

FIGURE 6-14
Absorption coefficient of 19 mm thick perforated fiberboard tiles on 25 mm battens. See text for description of *A*, *B*, and *C*. (Reprinted, by permission, from Reference 7.)

Perforated Covers Backed by Absorbent Materials

It is convenient to distinguish between two ways in which perforated covers are used. In the first case they are employed principally to protect the absorbent and to provide a suitable appearance, and they are intended to have a minimum effect on the absorption of the material underneath. Perforated covers also may be employed to control the properties of the

absorbent, the arrangement functioning as an array of Helmholtz resonators.

The effect of a perforated cover depends on its thickness and on the size
and spacing of the perforations. The perforations are usually circular, but
they are sometimes in the form of slits.

Perforated covers that are intended to be transparent to sound are
usually made of thin sheet metal, and this type of cover has been referred
to previously in the discussion of porous absorbents. The principal effect
of such covers is to reduce the absorption at high frequencies. The
frequency at which the reduction in efficiency becomes apparent, in the
case of circular perforations, can be roughly estimated from the formula

$$f = 1016 \frac{p}{d} \quad \text{Hz} \quad (6\text{-}10)$$

where d is the diameter of the perforations in mm, and p is the percentage
of open area. For a given percentage open area it is an advantage to
reduce the diameter of the perforations and to use a correspondingly larger
number.

When the area of the perforation is decreased, or the thickness of the
cover is increased, the facing has a greater effect on the absorption, and
the arrangement acts as a resonant absorber. The resonant frequency for a
perforated panel backed by a subdivided air space is given approximately
by

$$f = 5080 \sqrt{\frac{p}{l(t + 0.8d)}} \quad \text{Hz} \quad (6\text{-}11)$$

where l is the depth of the airspace, t the thickness of the cover, d the
diameter of the holes (all in mm) and p the percentage of open area.

When absorbent is introduced, the resonance frequency depends on the
properties of the absorbent and the formula is not generally applicable, but
it is useful as a guide to the order of magnitude of the resonance
frequency, and as an indication of the effect of the dimensions and spacing
of the perforations and the depth of the air space.

Figures 6-15 to 6-19 give a series of results that indicate the range of
absorbing properties obtained by using perforated panel absorbents. Figure 6-15 refers to a cover that provides almost complete transmission into
the absorbent, and Figure 6-16 shows the development of resonant absorption in the case of a thin sheet as the perforation is reduced. Figure 6-17
illustrates the use of a thick cover to obtain absorption at low frequencies.

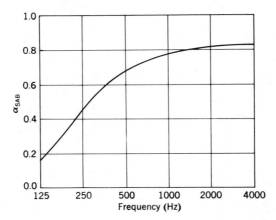

FIGURE 6-15
Absorption coefficient of 22-gauge perforated metal panels with 3.2 mm diameter holes, 23% open area, and backed by 38 mm thick absorbent. (Reprinted, by permission, from Reference 7.)

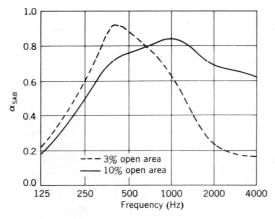

FIGURE 6-16
Absorption coefficient of 22-gauge perforated metal panels with 4.8 mm diameter holes, having 3 and 10% open area, and backed by 51 mm thick absorbent. (Reprinted, by permission, from Reference 7.)

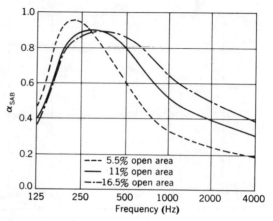

FIGURE 6-17
Absorption coefficient of 13 mm thick perforated plywood panels with 4.8 mm diameter holes, having 5.5, 11, and 16.5% open area and backed by 60 mm thick absorbent. (Reprinted, by permission, from Reference 7.)

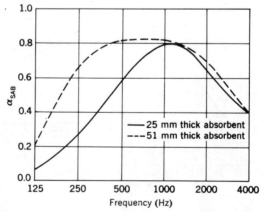

FIGURE 6-18
Absorption coefficient of 3 mm thick perforated hardwood panels with 4.8 mm diameter holes, having 10% open area and backed by 25 and 51 mm thick absorbent. (Reprinted, by permission, from Reference 7.)

192

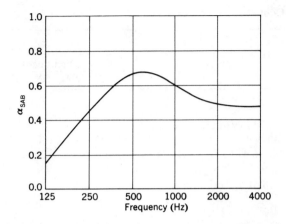

FIGURE 6-19
Absorption coefficient of 9.5 mm thick perforated plasterboard panels with 6.4 mm diameter holes, having 12% open area with tissue paper stuck to the back of the panels and on 51 mm battens. (Reprinted, by permission, from Reference 7.)

General Building Materials

Table 6-2 gives a brief list of common building materials that should prove useful in making estimates of room absorption.

TABLE 6-2
Sabine Absorption Coefficients for General Building Materials and Furnishings[a]

Materials	α_{SAB}					
	Octave band center frequencies (Hz)					
	125	250	500	1000	2000	4000
Brick, unglazed	.03	.03	.03	.04	.05	.07
Brick, unglazed, painted	.01	.01	.02	.02	.02	.03
Carpet						
3.2 mm pile height	.05	.05	.10	.20	.30	.40
6.4 mm pile height	.05	.10	.15	.30	.50	.55
4.8 mm combined pile and foam	.05	.10	.10	.30	.40	.50
7.9 mm combined pile and foam	.05	.15	.30	.40	.50	.60
Concrete block, painted	.10	.05	.06	.07	.09	.08

TABLE 6-2 (Cont'd)

Materials	α_{SAB}					
	Octave band center frequencies (Hz)					
	125	250	500	1000	2000	4000
Fabrics						
Light velour, 0.34 kg/m², hung straight, in contact with wall	.03	.04	.11	.17	.24	.35
Medium velour, 0.47 kg/m², draped to half area	.07	.31	.49	.75	.70	.60
Heavy velour, 0.61 kg/m², draped to half area	.14	.35	.55	.72	.70	.65
Floors						
Concrete or terrazzo	.01	.01	.01	.02	.02	.02
Linoleum, asphalt, rubber or cork tile on concrete	.02	.03	.03	.03	.03	.02
Wood	.15	.11	.10	.07	.06	.07
Wood parquet in asphalt on concrete	.04	.04	.07	.06	.06	.07
Glass						
6.4 mm, sealed, large panes	.05	.03	.02	.02	.03	.02
0.68 kg, operable windows (in closed condition)	.10	.05	.04	.03	.03	.03
Gypsum board, 13 mm nailed to 2×4's, 41 cm on center, painted	.10	.08	.05	.03	.03	.03
Marble on glazed tile	.01	.01	.01	.01	.02	.02
Plaster, gypsum or lime, rough finish or lath	.02	.03	.04	.05	.04	.03
Same, with smooth finish	.02	.02	.03	.04	.04	.03
Hardwood plywood paneling, 6.4 mm thick, wood frame	.58	.22	.07	.04	.03	.07
Water surface, as in a swimming pool	.01	.01	.01	.01	.02	.03
Wood roof decking, tongue-and-groove cedar	.24	.19	.14	.08	.13	.10

[a]From Reference 8, an annual publication, copyright 1975 by the Acoustical and Board Products Association, reproduced by permission. It is cautioned that this table lists approximate values which should be useful in making simple calculations of reverberation in rooms.

6.4 STEADY-STATE SOUND-PRESSURE LEVELS IN DIRECT AND REVERBERANT FIELDS

Direct Field

Assume that we have a sound source that is far removed from any walls and emits energy at a constant rate of W watts. The energy density [recall (1-7) and (1-8)] of the direct field is

$$D_o = \frac{WQ_\theta}{4\pi r^2 c} \qquad \text{W-sec}/\text{m}^3 \qquad (6\text{-}12)$$

where r is the distance from the source and Q_θ is the directivity factor of the source in a particular direction θ. The quantity Q_θ is a dimensionless number defined as the ratio of the intensity measured at an angle θ from an actual source to the intensity measured at the same distance from a nondirectional source, both sources radiating at the same power, W. In addition, values of Q_θ for a small nondirectional source located at the following positions in a large rectangular room are: (a) center, $Q_\theta = 1$; (b) center of one wall, $Q_\theta = 2$; (c) at the intersection of two walls halfway between floor and ceiling, $Q_\theta = 4$; and (d) at a trihedral corner, $Q_\theta = 8$.

Reverberant Field

The reverberant field differs from the direct field in that all waves have been reflected at least once, with different parts of the wavefronts having been incident (and reflected) at different angles on different elements of the boundaries of the enclosure. It can be shown that the steady-state energy density in the reverberant field is given by

$$D_R = \frac{4W}{cR_T} \qquad \text{W-sec}/\text{m}^3 \qquad (6\text{-}13)$$

where W is the power of the source and R_T is a room absorption constant corrected for the energy absorption of air, which is given by

$$R_T = \frac{S(\bar{\alpha}_{\text{ST}})_T}{1 - (\bar{\alpha}_{\text{ST}})_T} \qquad \text{m}^2 \qquad (6\text{-}14)$$

where

$$(\bar{\alpha}_{\text{ST}})_T = \bar{\alpha}_{\text{ST}} + \frac{4mV}{S} \qquad (6\text{-}15)$$

In (6-15) V is the volume of the enclosure, $\bar{\alpha}_{ST}$ is the average statistical absorption coefficient (recall p. 178) given by

$$\bar{\alpha}_{ST} = \frac{1}{S} \sum_{i=1}^{N} S_i (\alpha_{ST})_i \qquad (6\text{-}16)$$

where $(\alpha_{ST})_i$ is the statistical absorption coefficient corresponding to an area S_i, $S = \sum_{i=1}^{N} S_i$ is the total area, and m is the energy attenuation constant which depends strongly upon frequency and relative humidity. To a somewhat lesser degree, m is also temperature dependent. Values of m are given in Figure 6-20 and Table 6-3.

From (6-15) we see that when $\bar{\alpha}_{ST} \gg 4mV/S$; that is, when the absorption per reflection is large compared with air absorption between reflections, air absorption may be ignored. This is nearly always the case at frequencies of 1000 Hz or less, or in small rooms at all frequencies.

There is some controversy regarding (6-14). In Beranek[10] it is suggested that $\bar{\alpha}_{SAB}$ should be used. That is,

$$R_T = S (\bar{\alpha}_{SAB})_T = S\bar{\alpha}_{SAB} + 4mV \qquad \text{m}^2 \qquad (6\text{-}17)$$

where $\bar{\alpha}_{SAB}$ is given by (6-4). In practical problems they recommend (6-17) in preference to (6-14).

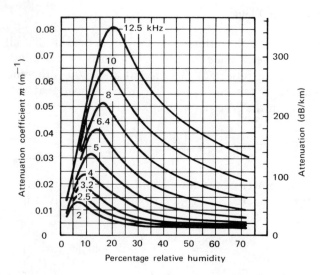

FIGURE 6-20

Coefficient m as a function of frequency and humidity at 20°C. (Reprinted, by permission, from C. M. Harris, "Absorption of Sound in Air in the Audio-Frequency Range," *J. Acoust. Soc. Am.*, Vol. 35, No. 1 (January 1963), pp. 11–17.)

TABLE 6-3
Coefficient $4m$ (m^{-1}) as a Function of Temperature and Humidity[a]

Relative humidity(%)	Temperature (°C)	Frequency (Hz) 2000	4000	6300	8000
30	15	0.0143	0.0486	0.1056	
	20	0.0119	0.0379	0.0840	0.136
	25	0.0114	0.0313	0.0685	
	30	0.0111	0.0281	0.0564	
50	15	0.0099	0.0286	0.0626	
	20	0.0096	0.0244	0.0503	0.086
	25	0.0095	0.0235	0.0444	
	30	0.0092	0.0233	0.0426	
70	15	0.0088	0.0223	0.0454	
	20	0.0085	0.0213	0.0399	0.060
	25	0.0084	0.0211	0.0388	
	30	0.0082	0.0207	0.0383	

[a] From *Noise and Vibration Control* by L. L. Beranek, Ed., Copyright 1971 by McGraw-Hill Book Company. Used with permission of McGraw-Hill Book Company.

As a final remark it is mentioned that the quantity $4V/S$ in the above equation is called the mean free path, which is defined as the average distance a sound wave travels in a room between reflections from the bounding surfaces. It can be derived theoretically and is the basis for the reverberation time formulas given on p. 207.

Total Steady-State Sound-Pressure Levels

At a point located a distance r from a single source the direct field is given by (6-12) and the reverberant field by (6-13). The total energy density is given by $D = D_0 + D_R$, or in terms of the mean-square sound pressure [recall (1-8)],

$$p_{\text{rms}}^2 = W\rho c \left[\frac{Q_\theta}{4\pi r^2} + \frac{4}{R_T} \right] \quad (\text{Pa})^2 \qquad (6\text{-}18)$$

The sound pressure level L_p is, [recall (1-20)],

$$L_p = L_w + 10\log_{10}\left(\frac{Q_\theta}{4\pi r^2} + \frac{4}{R_T} \right) \qquad \text{dB re 20 } \mu\text{Pa} \qquad (6\text{-}19)$$

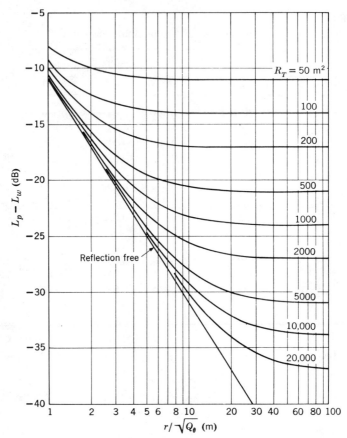

FIGURE 6-21
Difference between the sound-pressure level and the sound-power level in a room as a function of the room constant R_T, the distance from the source r, and the directivity factor Q_θ.

where L_w is the sound-power level in dB re 10^{-12} W. Equation 6-19 is shown in Figure 6-21.

From Figure 6-21 the relative importance of the contributions of the direct-and reverberant-field is evident. When $4/R_T$ is large compared to $Q_\theta/(4\pi r^2)$ the room absorption is an important factor in determining L_p. Conversely, when $4/R_T$ is small compared to $Q_\theta/(4\pi r^2)$ the room has a minimum effect on L_p and its value is very close to that obtained in a reflection-free (anechoic) environment. For this latter case L_p decreases 6 dB per doubling of distance and one can safely assume to be in the far field. However, in the former case it is seen that this 6 dB per doubling of distance is no longer true and one, therefore, is no longer in the far field. Furthermore, when $4/R_T$ is much greater than $Q_\theta/(4\pi r^2)$ (6-18) shows that p_{rms}^2 is proportional to the sound power. This provides the basis for determining the sound power radiated by a source in a reverberation room (see p. 211).

Example 6-2. A large rectangular room has a width of 24 m, a length of 36 m, and a height 10 m. In a 1/3-octave band it is found that the ceiling has a Sabine absorption coefficient of 0.3, the side walls, 0.4, the rear wall, 0.6, and the front wall and floor, 0.5. An omnidirectional sound source of 0.25 W is placed at the intersection of a wall and the floor. What is the sound-pressure level 9 m from the source?

Solution: The area of the floor or ceiling is 864 m², the side walls, 360 m², and the front or back wall, 240 m². The total area is 2928 m². The room constant is determined from (6-17) and (6-4). Thus

$$R_T = R = (864)(0.3) + (2)(360)(0.4) + (240)(0.6) + (240)(0.5)$$

$$+ (864)(0.5) = 1243.2 \text{ m}^2$$

The sound-power level is determined from (1-13). Thus

$$L_w = 10\log_{10} 0.25 + 120 = 114 \text{ dB re } 10^{-12} \text{ W}$$

The sound-pressure level is obtained from (6-19). Hence

$$L_p = 114 + 10\log_{10}\left[\frac{4}{(4\pi)(9)^2} + \frac{4}{1243.2}\right]$$

$$= 92.5 \text{ dB re } 20 \text{ }\mu\text{Pa}$$

Example 6-3. A source having a sound-power level L_w is placed in a room. The room originally has a room constant R_1. The absorbing properties of the room are changed so that the room constant is now R_2. Develop an expression to determine the change in the sound-pressure level at any position in the room caused by the change in the room constant.

Solution: The sound-pressure level associated with R_1 is

$$L_1 = L_w + 10\log_{10}\frac{Q}{4\pi r^2} + 10\log_{10}(1 + B_1) \qquad \text{dB re } 20 \text{ }\mu\text{Pa}$$

whereas that with R_2 is

$$L_2 = L_w + 10\log_{10}\frac{Q}{4\pi r^2} + 10\log_{10}(1 + B_2) \qquad \text{dB re } 20 \text{ }\mu\text{Pa}$$

where

$$B_j = \frac{16\pi r^2}{QR_j} \quad j = 1, 2$$

The change in sound-pressure level is

$$\Delta L = L_2 - L_1 = 10 \log_{10} \frac{1 + B_2}{1 + B_1}$$

From the above results it is seen that when $B_2 > B_1$, $\Delta L > 0$, indicating an increase from the original sound-pressure level. Conversely, when $B_1 > B_2$, $\Delta L < 0$ and the sound-pressure level decreases.

Example 6-4.   A rectangular room has a length of 15 m, a width of 9 m, and a height of 3 m. The walls, ceiling, and floor each have a Sabine absorption coefficient of 0.1. If, in a given frequency band, all walls and the ceiling can have their absorption coefficients increased to 0.8, what is the magnitude of the decrease in the sound-pressure level in the center of the room when the noise source is in a trihedral corner of the room.

Solution: The area of the four walls is 144 m² and that of the ceiling or floor 135 m². The total area is 414 m². Using (6-17) and (6-4) yields the room constant for the original absorption:

$$R_1 = (414)(0.1) = 41.4 \text{ m}^2$$

The room constant with the new absorption properties is

$$R_2 = (144)(0.8) + (135)(0.8) + (135)(0.1) = 236.7 \text{ m}^2$$

The distance from the source to the center of the room is

$$r = \sqrt{(7.5)^2 + (4.5)^2 + (1.5)^2} = 8.87 \text{ m}$$

The values of B_j, $j = 1, 2$ are

$$B_1 = \frac{(16\pi)(8.87)^2}{(8)(41.4)} = 11.95$$

$$B_2 = \frac{(16\pi)(8.87)^2}{(8)(236.7)} = 2.09$$

Using the expression for ΔL developed in Example 6-3 we have

$$\Delta L = 10 \log_{10} \frac{1 + 2.09}{1 + 11.95} = -6.2 \text{ dB}$$

Thus the increased absorption will bring a decrease of 6.2 dB in the sound-pressure level in the center of the room.

Example 6-5. Two sources are positioned in a room that has a constant R_T. One source has a directivity factor Q_1, a sound-power level L_{w1}, and is a distance r_1 from the point of interest. The other source has a directivity factor Q_2, a sound-power level L_{w2}, and is a distance r_2 from the point of interest. Develop an expression that relates the increase in the sound-pressure level in the room due to the second source.

Solution: Using (6-19) we have that each source individually generates the following sound-pressure levels:

$$L_1 = L_{w1} + 10\log_{10}(1 + A_1) + 10\log_{10}\frac{4}{R_T} \qquad \text{dB re 20 } \mu\text{Pa}$$

$$L_2 = L_{w2} + 10\log_{10}(1 + A_2) + 10\log_{10}\frac{4}{R_T} \qquad \text{dB re 20 } \mu\text{Pa}$$

where

$$A_j = \frac{Q_j R_T}{16\pi r_j^2} \qquad j = 1, 2$$

The total sound-pressure level L_T is obtained from (1-24). Thus

$$L_T = 10\log_{10}(10^{L_1/10} + 10^{L_2/10}) \qquad \text{dB re 20 } \mu\text{Pa}$$

The increase in the sound-pressure level ΔL is

$$\Delta L = L_T - L_1 = 10\log_{10}(10^{L_1/10} + 10^{L_2/10}) - L_1$$

$$= 10\log_{10}(1 + 10^{(L_2 - L_1)/10}) \qquad \text{dB}$$

Using the expressions for L_1 and L_2 yield

$$\Delta L = 10\log_{10}\left[1 + \left(\frac{1 + A_2}{1 + A_1}\right)10^{(L_{w2} - L_{w1})/10}\right] \qquad \text{dB}$$

If W_1 and W_2 are the acoustic power of each source such that

$$L_{w_j} = 10\log_{10}W_j + 120 \qquad \text{dB re } 10^{-12} \text{ W} \qquad (j = 1, 2)$$

then ΔL can be rewritten as

$$\Delta L = 10\log_{10}\left[1 + \left(\frac{1 + A_2}{1 + A_1}\right)\left(\frac{W_2}{W_1}\right)\right] \qquad \text{dB}$$

Example 6-6.   Two omnidirectional sources are placed in a room having a constant $R_T = 400$ m^2. One source has a sound power of 0.1 W and is located 6 m from the center of the room. The second source has an acoustic power of 0.05 W and is located 3.4 m from the room center. What is the increase in the sound-pressure level in the center of the room due to the second source if it is assumed that $Q_1 = Q_2 = 1$.

Solution: From Example 6-5 we have

$$A_1 = \frac{400}{(16\pi)(6)^2} = 0.221$$

$$A_2 = \frac{400}{(16\pi)(3.4)^2} = 0.688$$

and, therefore, the increase is

$$\Delta L = 10\log_{10}\left[1 + \left(\frac{1.688}{1.221}\right)\left(\frac{0.05}{0.10}\right)\right] = 2.3 \text{ dB}$$

Example 6-7.   A source with sound-power level L_{w1} is placed in a room having a constant R_T. It is found necessary to place another source in the room that has a sound-power level L_{w2}. The distance from the first source to the point of interest is r_1, and that of the second source, r_2. Develop an expression from which the value of a new room constant can be determined such that there is no increase in the sound-pressure level at the point of interest.

Solution: The sound-pressure level caused by the first source in the original room is, from (6-19),

$$L_1 = L_{W1} + 10\log_{10}\left[\frac{Q_1}{4\pi r_1^2} + \frac{4}{R_T}\right] \qquad \text{dB re 20 } \mu\text{Pa}$$

The total sound pressure level L_T' from both sources in the room with the new room constant R_T', using the results from Example 6-5, is

$$L_T' = L_{W1} + 10\log_{10}\left[\frac{Q_1}{4\pi r_1^2} + \frac{4}{R_T'}\right]$$

$$+ 10\log_{10}\left[1 + \left(\frac{W_2}{W_1}\right)\left(\frac{1 + A_2'}{1 + A_1'}\right)\right] \qquad \text{dB re 20 } \mu\text{Pa}$$

where

$$A_j' = \frac{Q_j R_T'}{16\pi r_j^2} \quad j = 1, 2$$

The difference in the sound-pressure level in the original room with one source and the new room with two sources is the difference between L_T' and L_1. Thus

$$\Delta L = L_T' - L_1 = 10\log_{10}\left\{\left[\frac{R_T}{R_T'}\left(1 + \frac{W_2}{W_1}\right) + \frac{R_T}{16\pi}\left(\frac{Q_1}{r_1^2} + \frac{W_2}{W_1}\frac{Q_2}{r_2^2}\right)\right]\Bigg/\right.$$
$$\left.\left[1 + \frac{R_T Q_1}{16\pi r_1^2}\right]\right\} \quad \text{dB}$$

When $\Delta L > 0$ there is an increase over that sound-pressure level at the point of interest due to a single source L_{w1} in the room with constant R_T. When $\Delta L < 0$ there is a decrease. When $\Delta L = 0$ both L_1 and L_T' are equal, that is, the room absorption increases such that there is no net increase in the total sound-pressure level.

For either of these three cases we can solve for R_T' explicitly. Thus

$$R_T' = R_T\frac{A}{B} = R_T\left\{\left(1 + \frac{W_2}{W_1}\right)\Bigg/\left[\frac{Q_1 R_T}{16\pi r_1^2}(10^{\Delta L/10} - 1)\right.\right.$$
$$\left.\left. + 10^{\Delta L/10} - \left(\frac{W_2}{W_1}\right)\frac{R_T Q_2}{16\pi r_2^2}\right]\right\} \geqslant R_T > 0$$

For the above expression for R_T' to make sense in a practical situation, $A/B \geqslant 1$ and $R_T'/S \leqslant \bar{a} < 1$, where S is the total area of the room.

Example 6-8. A room that is 14 m long, 11 m wide, and 3.5 m high has a constant $R_T = 120$ m². On the floor in one corner of the room, an omni-directional source of 2.0 W is placed. Another source also having 2.0 W of power is going to be placed on the floor in the diagonally opposite corner of the room. What must be the value of the new room constant so that there will not be an increase in the sound-pressure level in the center of the room.

Solution: The distance from each source to the center of the room is $r_1 = r_2 = [(7)^2 + (5.5)^2 + (1.75)^2]^{1/2} = 9.07$ m. The new room constant, using the results of Example 6-7 and the fact that $\Delta L = 0$, is

$$R'_T = 120\left(1 + \frac{2}{2}\right)\left[1 - \left(\frac{2}{2}\right)\frac{(120)(8)}{(16\pi)(9.07)^2}\right]^{-1} = 312.6 \ m^2$$

The total room surface area is 483 m^2. Thus the average absorption coefficient must be increased from $\bar{\alpha} = 120/483 = 0.25$ to $\bar{\alpha}' = 312.6/483 = 0.65$ so that the increase in sound-pressure level due to the second source is nullified.

Example 6-9. A room that is 9 m long, 4.5 m wide, and 3 m high has an air conditioning duct outlet in the center of its ceiling. The octave band power levels of the outlet are given in the solution to Example 5-9. Determine the average absorption coefficient for the room surfaces in each octave band such that at a distance of 2.5 m away from the duct outlet the sound-pressure level does not exceed PNC-30. (See Figure 3-9.)

Solution: Using (6-17) and (6-19) yields

$$1 \geqslant \bar{\alpha}_{SAB} = \frac{4}{S}\left[10^{(L_p - L_w)/10} - \frac{Q}{4\pi r^2}\right]^{-1} > 0$$

where S is the total area of the room, L_p is the sound-pressure level of the PNC-30 contour given in Figure 3-9, and L_w is the sound-power level of the duct outlet given in the solution to Example 5-9. Since the duct outlet is flush with the ceiling $Q_\theta = 2$. The total area is 162 m^2. The results are summarized in Table 6-4. It can be seen from Table 6-4 that at 500 Hz the PNC-30 curve cannot be attained.

TABLE 6-4
Tabulation of the Solution to Example 6-9

Octave band center Frequency (Hz)	Sound-pressure level of PNC-30 (dB re 20 μPa)	Sound-power level of duct outlet (dB re 10^{-12} W)	Computed $\bar{\alpha}_{SAB}$
125	46	51.5	0.096
250	41	50.5	0.29
500	34	48.5	2.46
1000	30	42.5	0.80
2000	25	34.5	0.29
4000	23	17.5	a
8000	23	0	a

[a] Negligible.

6.5 ROOM REVERBERATION

Reverberation Time—Introduction

The reverberation time of an enclosure is defined as the time required for the sound-pressure level to decrease 60 dB. This time corresponds approximately to that required for a sound to diminish from a fairly loud level in a room to the threshold of audibility.

In continuing speech and music it is not possible for a listener to perceive a decay of 60 dB, since the latter portion of the decay is masked by following sounds. The subjective impression of reverberation time is thus found to depend on the initial decay rates. Although classical theory predicts an exponential decay (which is represented as a straight line on a logarithmic decay curve), deviations from this are found in practice.

The concept of reverberation is very closely tied to the idea of echoes. The direct sound coming from the source to the listener without having been reflected by any of the bounding surfaces plays an important part in the subjective appraisal of the acoustics of a room. Under steady-state conditions the level of the direct sound falling off with distance from the source is very quickly less than the reverberant sound level, but with transient sound the direct path is the shortest, and sound by this path is the first to reach the listener. Sounds reflected from nearby surfaces will follow with a short time delay depending on the extra path travelled. (An extra path difference of 35 cm corresponds approximately to a time delay of 1 msec.) The sound that reaches the listener first determines the apparent direction from which the sound comes even though the delayed sound, in this case from reflections, may actually be more intense.

A sense of "volume" of the room appears to be an important subjective criterion for music, and is related to the perception of reflected sound coming from many different directions. Whether or not a reflection will be perceived depends on the degree of masking present—which in turn depends on the relative levels and delays of the direct and other reflected sounds. Some work has been carried out in an attempt to determine masking levels for reflected sounds coming from different directions. It appears that a sound originating within a small angle of another sound direction will tend to be masked. In addition, side-wall reflections received within the body of the audience will suffer considerable attenuation through diffraction and absorption at grazing incidence, and will tend to be masked by reflections coming from the ceiling, which are less attenuated. Thus a listener will have difficulty in perceiving lateral reflections unless they arrive before those coming from overhead. The short-path overhead reflections themselves are also probably masked to some extent by the direct sound, since localization in the vertical plane tends to be very poor and greatest masking occurs from sounds coming from similar directions laterally.

From this discussion it is evident that a listener seated away from lateral reflectors will have difficulty perceiving the directional origin of reflected sounds, and thus will have little auditory sense of the size and shape of the room in which he is sitting. It has been suggested that a spatial impression of a sound field will be perceived if the sound comes from four or five distinctly different horizontal directions—provided that these signals are not coherent. In other words, a pleasing quality of reverberation is not achieved where the sound field becomes essentially isotropic after many reflections and there is great diffusion of the sound, but rather where some directionality is maintained, with sufficient diffusion to prevent echoing or focusing. Conditions are also unfavorable when a large fraction of the sound energy comes from overhead at large angles.

Typical values for reverberation times for various types of rooms and various types of usage are shown in Figure 6-22. It is generally accepted that, in addition to an optimum reverberation time in the midfrequency range, the reverberation time as a function of frequency should have a suitable characteristic throughout the audible range. Experience shows that

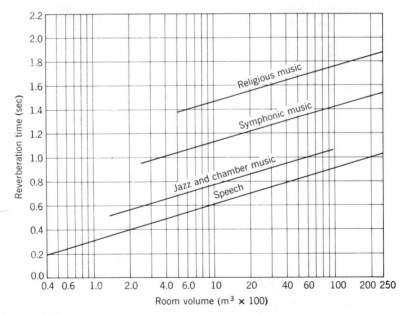

FIGURE 6-22
Optimum reverberation times for rooms in the frequency range 500–1000 Hz.

there is usually a progressive shortening of the reverberation times with higher frequencies because of sound absorption by air, audience, and normal furnishings of a hall. Furthermore, a rising reverberation time characteristic toward the lower frequencies is highly desirable. This low-frequency reverberation gives fullness of tone to music and body to speech, and is needed both because of the reduced sensitivity of the ear at lower frequencies and to compensate for the reduced output of many sound sources in this range. This increase in reverberation time at 100 Hz is 1.4–1.75 times the reverberation time at 500–1000 Hz.

Reverberation Time Formulas

The reverberation time T_{60} is given by the following relation*

$$T_{60} = \frac{55.3}{c}\frac{V}{Sa_T} = 0.161\frac{V}{Sa_T} \quad \text{sec} \tag{6-20}$$

where V is the volume (m^3) and S is the surface area (m^2). The quantity a_T is given by either of the following expressions:

Sabine

$$a_T = (\bar{\alpha}_{SAB})_T \tag{6-21}$$

or

Eyring

$$a_T = -2.30\log_{10}[1 - (\bar{\alpha}_{ST})_T] \tag{6-22}$$

where $(\bar{\alpha}_{SAB})_T$ is given by (6-17) and $(\bar{\alpha}_{ST})_T$ by (6-15). In Beranek it is recommended that that (6-21) be used for most engineering design work.

Fitzroy[14] modified (6-20) and (6-22) to obtain good agreement between theory and experiment for rooms in which pairs of parallel surfaces have widely varying average absorption coefficients. The Fitzroy relation is

$$T_{60} = \frac{0.161V}{S}\left[\frac{S_v}{a_v} + \frac{S_t}{a_t} + \frac{S_l}{a_l}\right] \quad \text{sec} \tag{6-23}$$

where the subscripts v, t, and l denote to the floor-ceiling, the side-walls, and end walls, respectively. For example, S_v is the area of the floor and ceiling and a_v is the corresponding weighted coefficient given by (6-22) with $(\bar{\alpha}_{ST})_T$ determined by the properties of the floor and ceiling. Equation (6-23) could also be rewritten using a_v, and so forth, defined by (6-21).

---

*There are numerous forms for (6-20). For a detailed discussion see References 11–13 and pp. 233–243 of Reference 10.

Example 6-10. Consider the following three rooms, for which the dimensions and corresponding statistical absorption coefficients are summarized below. Determine reverberation times using the Eyring formula and the Fitzroy-Eyring formula. Compare these values with the experimentally determined ones.

Room	Room dimensions (m)			$\bar{\alpha}_{ST}$		
Number	Height	Width	Length	$\bar{\alpha}_v$	$\bar{\alpha}_t$	$\bar{\alpha}_l$
1	3.66	9.14	11.0	0.39	0.03	0.03
2	4.42	16.8	17.4	0.40	0.14	0.03
3	6.10	16.5	24.7	0.15	0.20	0.29

Solution: The results are summarized below.

Room	Room volume	Total area	Area of parallel surfaces (m²)			Absorption constants				
						(6-22)			(6-16)	(6-22)
Number	(m³)	(m²)	S_v	S_t	S_l	a_v	a_t	a_l	$\bar{\alpha}_{ST}$	a_T
1	368	348.5	201	80.5	66.9	0.49	0.03	0.03	0.24	0.27
2	1292	887	584.6	153.8	148.5	0.51	0.15	0.03	0.30	0.36
3	2486	1318	815	301.3	201.3	0.16	0.22	0.34	0.34	0.20

	Reverberation times (sec)		
Room	(6-20) and (6-22)	(6-23)	Experimental[14]
number	T_E	T_F	T_{EX}
1	0.63	2.59	2.55
2	0.67	1.78	1.70
3	1.51	1.62	1.57

Determination of the Sabine Absorption Coefficient[15, 16]

Using (6-20) and (6-21) we can determine $\bar{\alpha}_{SAB}$ for a given material. Consider first an empty reverberation room. (See p. 211 for the properties of a reverberation room.) From (6-20) we have

$$T_1 = 0.161 \frac{V}{S(\bar{\alpha}_{SAB})_1} \qquad \text{sec} \qquad (6\text{-}24)$$

from which we can determine $(\bar{\alpha}_{SAB})_1$ from the measurement of the reverberation time of the room. We now introduce on one wall a material of area S_A whose absorption is denoted $(\bar{\alpha}_{SAB})_2$ and determine the new reverberation time T_2, which is related to $(\bar{\alpha}_{SAB})_2$ by

$$T_2 = \frac{0.161\,V}{(S - S_A)(\bar{\alpha}_{SAB})_1 + S_A\,(\bar{\alpha}_{SAB})_2} \qquad \text{sec} \qquad (6\text{-}25)$$

Combining (6-24) and (6-25) yields

$$(\bar{\alpha}_{SAB})_2 = (\bar{\alpha}_{SAB})_1 + \frac{0.161\,V}{S_A}\left(\frac{1}{T_2} - \frac{1}{T_1}\right) \qquad (6\text{-}26)$$

It is noted that T_1 is always greater than T_2.

6.6 NUMBER OF MODES AND THE MODAL DENSITY OF A ROOM

Consider a rectangular enclosure with dimensions L_x, L_y, and L_z. If the walls of this enclosure are rigid it can be shown that the pressure at any point in the room is proportional to

$$p \sim \sum_{n=0}^{\infty} \sum_{m=0}^{\infty} \sum_{l=0}^{\infty} A_{nmk} \cos\left(\frac{n\pi x}{L_x}\right)\cos\left(\frac{m\pi y}{L_y}\right)\cos\left(\frac{l\pi z}{L_z}\right) \qquad (6\text{-}27)$$

wherein we have chosen the origin of the coordinate system to coincide with a corner of the room. From (6-27) we see that p is a maximum at any of the eight corners of the enclosure. Thus to excite the maximum number of room modes, one should place a source in the corner of the room. The room modes, given by (6-27), occur at the resonant frequencies of the room given by*

$$f = \frac{c}{2}\left[\left(\frac{n}{L_x}\right)^2 + \left(\frac{m}{L_y}\right)^2 + \left(\frac{l}{L_z}\right)^2\right]^{1/2} \qquad \text{Hz} \qquad (6\text{-}28)$$

We know that there is a value of f for every combination of the independent variable integers n, m, and l. Equation 6-28 can therefore be represented by the rectangular lattice shown in Figure 6-23. Each intersection shown by a dot in Figure 6-23 represents one solution to (6-28) and the appropriate frequency is equal to the length of the vector drawn from the origin to the intersection. The number of resonances N below a certain frequency f is equal to the number of intersections contained within the volume bounded by the three coordinate planes and a spherical surface of

*See Reference 4, pp. 582–599.

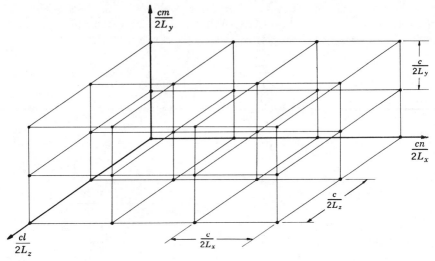

FIGURE 6-23
Graphical representation of (6-29); $m, n, k = 1, 2, 3, \ldots$.

radius f centered at the origin. Since there are $8V/c^3$ lattice cells per unit volume of frequency space ($V = L_x L_y L_z$), there will be on the average $(8V/c^3)(\pi f^3/6)$ normal modes having frequency equal to or less than f (the factor $\pi f^3/6$ being the volume of an eighth of a sphere of radius f). Thus

$$N = \frac{4\pi V f^3}{3c^3} \tag{6-29}$$

We can separate the standing waves into three categories: axial waves (for which two of the integers are zero), tangential waves, (for which one of the integers is zero), and oblique waves (for which none of the integers are zero). It can be shown that the total number of standing waves of all classes which have frequencies less than f is

$$N = \frac{4\pi f^3 V}{3c^3} + \frac{\pi f^2 A}{4c^2} + \frac{fL}{8c} \tag{6-30}$$

where $A = 2(L_x L_y + L_x L_z + L_y L_z)$ and $L = 4(L_x + L_y + L_z)$. We see that A is the total wall area and L is the sum of the lengths of all of the edges of the room.

The number of standing waves or modes with frequencies in a band of width Δf is obtained by differentiating (6-30). Thus

$$\Delta N = \left[\frac{4\pi f^2 V}{c^3} + \frac{\pi f A}{2c^2} + \frac{L}{8c} \right] \Delta f \tag{6-31}$$

The modal density, the number of modes per bandwidth Δf, is $\Delta N/\Delta f$. Conversely, the mean frequency spacing between resonance peaks is approximately $\Delta f/\Delta N$. It can be shown that for high frequencies, regardless of the shape of the room, the first term of (6-31) predominates and can be used to determine the number of resonance frequencies.

We notice from (6-31) that the average number of allowed frequencies in a band increases with the square of the frequency at higher frequencies. If we assume that the average intensity of sound in a room (for a constant output source) is proportional to the number of standing waves that carry the sound, the intensity in the room increases as the square of the frequency, for high frequencies. It can be shown that the power output into free space from a simple source is proportional to f^2. Therefore, the power transmitted from source to receiver in a room varies, on the average, with frequency, as it does in the open; but superimposed on the smooth rise are fluctuations resulting from the fluctuations of the number of standing waves in the frequency band of the driver. These irregularities of response are more pronounced the more symmetric the shape of the room, or the narrower the frequency band of the sound source.

6.7 SPECIAL TEST CHAMBERS

Introduction

As can be inferred from (6-19), the total acoustic power radiated by a source is frequently a more useful quantity than the acoustic pressure because the power is independent (in many instances) of the environment into which it is placed; the sound pressure, on the other hand, is not. There are two limiting cases of (6-19): (1) the room is perfectly absorbing and (2) the room is perfectly reflecting. The former constitutes an anechoic (without echo) room and the latter a reverberation room. Both of these rooms can be used to determine the sound power of a source.[17, 18] The anechoic room can additionally be used to determine the directivity of the source. The reverberation room can also be used to determine Sabine absorption coefficients (recall p. 208) and the transmission loss and impact isolation[19–21] of panels (see Section 7.4). The details of the procedures to perform these measurements are left to the references cited, and only the properties of the rooms themselves are discussed in the next sections.

Reverberation Room

The determination of the sound power radiated by the source of sound in a reverberation chamber is based on the premise that the sound field is diffuse. This diffusivity requirement is also necessary for absorption and transmission loss experiments. In a diffuse sound field the mean-square sound pressure in the chamber is simply related to the sound power radiated by the source. In practice a diffuse field can be established[22] in rooms of good design at frequencies above $f_c \approx 2000 \, (T_{60}/V)^{1/2}$ where T_{60} is the reverberation time in seconds and V is the room volume in cubic meters. This equation must be solved in an iterative fashion since T_{60} is also a function of frequency. If the reverberation room is to be used at frequencies below f_c other criteria must be satisfied as detailed in References 17 and 18.

For sound power, transmission loss, and absorption measurements it is recommended that the volume of the room(s) be at least 200 m^3, the smallest dimension of the room be at least one wavelength and preferably more than two wavelengths of the center frequency of the lowest 1/3-octave band at which measurements are to be made, and that the largest dimension (the major diagonal for a rectangular room) be less than 1.9 $(V)^{1/3}$. If the room is rectangular the ratio of any two dimensions should be such that no two modes have the same natural frequency [recall (6-28)]. The proportions that are frequently used are $1:2^{1/3}:4^{1/3}$.

For absorption measurements the empty room absorption should be less than 0.06. For sound-power measurements it is permissible for the absorption to increase to 0.16 in the lowest 1/3-octave band.

Anechoic Chamber

An anechoic chamber is a room, which over a frequency range, very closely approximates a free field environment, that is, one in which sound radiates from the source with none of its sound being reflected back. All the surfaces of the chamber are lined with absorbant wedges so that 99% of the incident energy is absorbed over the usable frequency range. The most common method of assessing anechoic chambers is the inverse distance test. The inverse distance test is based on the fact that, in free space, the sound pressure due to a point source radiating spherical waves of equal intensity in all directions is inversely proportional to the distance from the source [recall (1-4)]. Any deviation from this theoretical relationship is an indication of non-uniformity of the sound field, that is, departure from free space conditions. Some of the details of wedge design are given by Koidan et al.[23] and those considering the overall design of a large anechoic room by Ingerslev et al.[24]

REFERENCES

1. "Impedance and Absorption of Acoustical Materials by the Tube Method," ASTM C-384-58, American Society for Testing and Materials, Philadelphia, Pa.

2. J. G. Powell and J. J. Van Houten, "A Tone-Burst Technique of Sound-Absorption Measurement," *J. Acoust. Soc. Am.*, Vol. 48, No. 6 (1970), pp. 1299–1303.

3. P. Dubout and W. Davern, "Calculation of the Statistical Absorption Coefficient from Acoustic Impedance Tube Measurements," *Acustica*, Vol. 9 (1959), pp. 15–16.

4. P. M. Morse and K. U. Ingard, *Theoretical Acoustics*, McGraw-Hill, New York (1968), p. 580.

5. D. Olynyk and T. D. Northwood, "Comparison of Reverberation Room and Impedance Tube Absorption Measurements," *J. Acoust. Soc. Am.*, Vol. 36, No. 11 (Nov. 1964), pp. 2171–2174.

6. M. E. Delany and E. N. Bazley, "Acoustical Properties of Fibrous Absorbent Materials," *Appl. Acoust.* (3) (1970), pp. 105–116.

7. E. J. Evans and E. N. Bazley, "Sound Absorbing Materials," National Physical Laboratory, London; Her Majesty's Stationary Office, 1960 (4th impression 1968).

8. "Performance Data—Architectural Acoustical Materials," Annual Bulletin, Acoustical and Board Products Association, Park Ridge, Illinois.

9. W. E. Purcell, "Compendium of Materials for Noise Control," Final Technical Report, IITRI Project J6285, Engineering Mechanics Division, IIT Research Institute, Chicago, Ill. (April 1974). (Prepared for Department of Health, Education and Welfare, National Institute for Occupational Safety and Health, Cincinnati, Ohio.)

10. L. L. Beranek, Ed., *Noise and Vibration Control*, McGraw Hill, New York (1971), p. 228.

11. R. W. Young, "Sabine Reverberation Equation and Sound Power Calculations," *J. Acoust. Soc. Am.*, Vol. 31, No. 7 (July 1959), pp. 912–921.

12. T. F. W. Embleton, "Absorption Coefficients of Surfaces Calculated from Decaying Sound Fields," *J. Acoust. Soc. Am.*, Vol. 50, No. 3 (1971), pp. 801–811.

13. M. C. Gomperts, "Do the Classical Reverberation Formulae Still Have a Right for Existence?," *Acustica*, Vol. 16 (1965/66), pp. 254–268.

14. D. Fitzroy, "Reverberation Formula Which Seems to be More Accurate with Nonuniform Distribution of Absorption," *J. Acoust. Soc. Am.*, Vol. 31, No. 7 (July 1959), pp. 893–897.

15. "Sound Absorption of Acoustical Materials in Reverberation Rooms," ASTM C423-66 (revised June 1970), American Society of Testing Materials, Philadelphia, Pa. (Same as ANSI S1.7-1970).

16. C. W. Kosten, "International Comparison Measurements in the Reverberation Room," *Acustica*, Vol. 10 (1960), pp. 400–411.

17. "Method for the Physical Measurement of Sound," ANSI S1.2-1962, American National Standards Institute, New York.

18. "Methods for the Determination of Sound Power Levels of Small Sources in Reverberation Rooms," ANSI S1.21-1972, American National Standards Institute, New York.

19. "Laboratory Measurement of Airborne Sound Transmission Loss of Building Partitions," ASTM E90-70, American Society of Testing and Materials, Philadelphia, Pa.

20. "Measurement of Airborne Sound Insulation in Buildings," ASTM E336-71, American Society of Testing and Materials, Philadelphia, Pa.

21. "Tentative Method of Laboratory Measurement of Impact Sound Transmission Through Floor-Ceiling Assemblies using the Tapping Machine," ASTM E492-73T, American Society of Testing and Materials, Philadelphia, Pa.

22. D. Lubman, "Precision of Reverberant Sound Power Measurements," *J. Acoust. Soc. Am.*, Vol. 52, No. 2 (August 1974), pp. 523–533.

23. W. Koidan, G. R. Hruska, and M. A. Pickett, "Wedge Design for National Bureau of Standards Anechoic Chamber," *J. Acoust. Soc. Am.*, Vol. 52, No. 4 (1972), pp. 1071–1076.

24. F. Ingerslev, O. J. Pedersen, P. K. Moller, and S. Kristensen, "New Rooms for Acoustic Measurements at the Danish Technical University," *Acustica*, Vol. 19 (1967/68), pp. 185–199.

7

TECHNIQUES FOR
THE REDUCTION
OF SOUND
AND VIBRATION

7.1 ACOUSTIC SILENCERS AND MUFFLERS[1-3]

Introduction

In noise-control work, the need for an acoustic filter is usually encountered when it becomes necessary to transport air or other gases from one place to another through a duct or a tube. If one of the requirements of the duct installation is that transmission of noise through the duct be held to a minimum, some acoustic device must be inserted in the duct. Two general principles are employed in obtaining the filtering action: absorption and reflection. The former principle yields a dissipative-type muffler and the latter a reactive-type muffler.

A dissipative muffler is one whose acoustical performance is determined mainly by the presence of sound-absorbing materials. A reactive muffler is one whose acoustical performance is determined mainly by its geometrical shape. When a sound wave traveling through a duct arrives at a discontinuity where the acoustical impedance is either much higher or much lower than the characteristic impedance of the duct, only a small portion of the acoustical energy can flow through the discontinuity. The rest of the energy goes into a reflected wave that originates at the discontinuity and travels back toward the source. Thus the transmission of sound energy can be reduced by inserting appropriate discontinuities in the duct, even though these discontinuities may not actually absorb any of the energy. Reflective acoustical filters are most effective at low frequencies in contrast to sound-absorptive devices, which are usually most effective at high frequencies.

Several terms used to describe the efficacy of acoustic silencing devices are now introduced.

Insertion Loss: The difference in dB between two sound-pressure levels (or power or intensity levels) that are measured at the same point in space before and after a muffler is inserted between the measurement point and the noise source. It is important to recognize that the insertion loss is not a unique property of the filter; it depends also on the source and terminating impedance, and the location of the point at which the insertion loss is determined. As a result, it is not practical to present general curves for the insertion loss of filters of various types; separate calculations are normally required for individual cases.

Transmission Loss (TL): The relationship between the energy in the incident wave at the inlet and the energy in the transmitted wave at the outlet. It is usually expressed in dB. The TL does not depend on the source impedance, although it does depend on the terminating impedance. The TL curves are usually presented for the special case of a reflection-free termination.

Noise Reduction: The difference, in dB, between the sound-pressure levels measured at the input of a muffler and at its output.

Attenuation Level: The decrease of sound power, in dB, between two points in an acoustical system. It is a useful quantity for describing wave propagation in lined ducts where the acoustically effective material is continuously and uniformly distributed along the direction in which energy flows. In this case, the attenuation is measured by determining the decrease in sound-pressure level per unit length of duct.

Transmission Loss of Reactive Filters Terminated Without Reflection

Series Filters

Single Expansion Chamber: For this filter, shown in Figure 7-1a, the transmission loss is

$$TL = 10\log_{10}\left[1 + 0.25\,(m - m^{-1})^2 \sin^2 kL \right] \qquad dB \qquad (7\text{-}1)$$

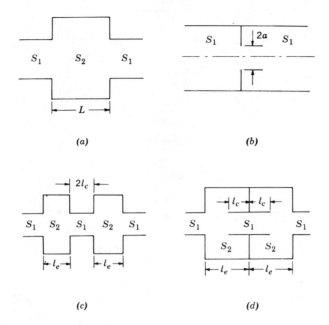

FIGURE 7-1
Schematic diagrams of various series filters: (*a*) single chamber (S_1 and S_2 are the cross-sectional areas); (*b*) orifice in duct (*a* = radius of the orifice); (*c*) double-expansion chamber with external connecting tube; (*d*) double-expansion chamber with internal connecting tube.

where $m = S_2/S_1$ and $k = 2\pi/\lambda = 2\pi f/c$ is the wave number. Equation 7-1 is plotted in Figure 7-2. Equation 7-1 (and all subsequent equations) is valid when $d/\lambda \leqslant 1.22$ or $f \leqslant 1.22\, c/d$ where d is the largest dimension of the cross-section, λ the wavelength, c the speed of sound, and f the frequency. This is the upper limit for the plane wave assumption in ducts.

The abrupt change in area need not be the case. The different cross-sections may be connected by a steep tapered section (truncated cone). If the connection (taper) is steep enough (7-1) is still valid. However, a long, slender taper would act as a horn and reduce the filter effectiveness severely at high frequencies. This is shown in Figure 7-3.

Orifice in Duct: For this filter, shown in Figure 7-1*b*, the transmission loss is

$$TL = 10\log_{10}\left[1 + (kS/4a)^2\right] \qquad dB \qquad (7\text{-}2)$$

Equation 7-2 is plotted in Figure 7-4.

Double-Expansion Chamber with External Connecting Tube: For this filter,

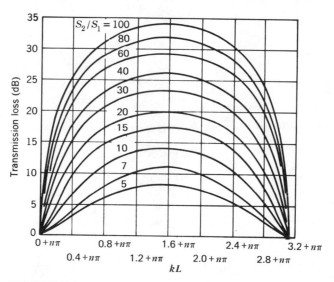

FIGURE 7-2
Transmission loss of a single expansion chamber ($n = 0, 1, 2, \ldots$).

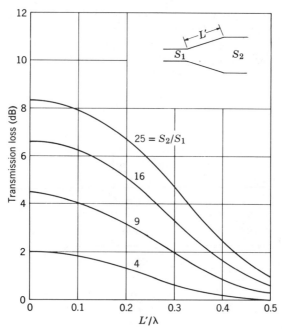

FIGURE 7-3
Transmission loss of a truncated cone.

218

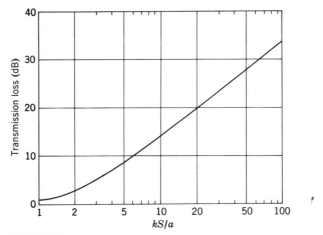

FIGURE 7-4
Transmission loss of an orifice in a duct.

shown in Figure 7-1c, the transmission loss is

$$TL = 10 \log_{10} [A_1^2 + B_1^2] \qquad dB \qquad (7\text{-}3)$$

where

$$A_1 = \frac{1}{16m^2} \left\{ 4m(m+1)^2 \cos 2k(l_e + l_c) - 4m(m-1)^2 \cos 2k(l_e - l_c) \right\}$$

$$B_1 = \frac{1}{16m^2} \left\{ 2(m^2+1)(m+1)^2 \sin 2k(l_e + l_c) - 4(m^2-1)^2 \sin 2kl_c \right.$$

$$\left. - 2(m^2+1)(m-1)^2 \sin 2k(l_e - l_c) \right\}$$

This equation is shown in Figure 7-5 for several combinations of parameters. We note that the transmission loss is higher than that of a single chamber. A low-frequency pass region is introduced as a result of resonance between the connecting tube and the chambers. When the connecting tube is lengthened, the lower cut-off frequency is reduced. An approximate equation for this cut-off frequency is

$$f_c \approx \frac{c}{2\pi} \left[ml_e l_c + \frac{l_e}{3}(l_e - l_c) \right]^{-1/2} \qquad Hz \qquad (7\text{-}4)$$

The maximum transmission loss in the first stop band above this cut-off frequency increases as the length of the connecting tube is increased.

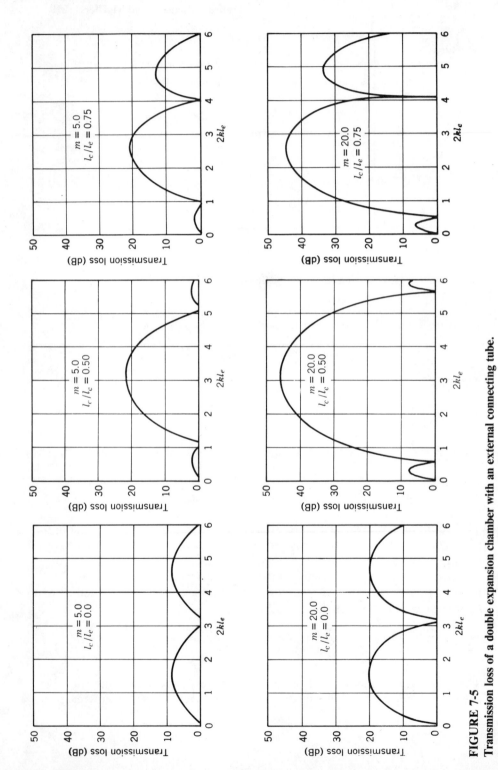

FIGURE 7-5
Transmission loss of a double expansion chamber with an external connecting tube.

However, the regions of low transmission loss have greater bandwidth when these connecting tubes are long.

Double-Expansion Chamber with Internal Connecting Tube: For this filter, shown in Figure 7-1*d*, the transmission loss is

$$TL = 10\log_{10}\left[A_2^2 + B_2^2\right] \qquad dB \qquad (7-5)$$

where

$$A_2 = \cos 2kl_e - (m-1)\sin 2kl_e \tan kl_c$$

$$B_2 = \frac{1}{2m}\left[(m^2+1)\sin 2kl_e + (m-1)\tan kl_c \left\{(m^2+1)\cos 2kl_e \right.\right.$$

$$\left.\left. -(m^2-1)\right\}\right]$$

These results are shown in Figure 7-6 for several combinations of parameters. The low-frequency pass region is again present, and the cut-off frequency is given by (7-4). It should be realized from Figures 7-1*c* and 7-1*d* that (7-3) equals (7-5) when $l_c = 0$.

Side-Branch (Parallel) Filters

Orifice: For this filter, shown in Figure 7-7*a*, the transmission loss is

$$TL = 10\log_{10}\left[1 + \left(\frac{S_0}{2kL'S}\right)^2\right] \qquad dB \qquad (7-6)$$

where $S_0 = \pi a^2$ is the area of the orifice and

$$L' = L_0 + 1.6a \qquad (7-7)$$

To arrive at (7-6) it was assumed that $L' \ll \lambda$. When this latter assumption no longer holds the orifice must be treated as a tube, as shown below. It is seen that the orifice acts as a high-pass filter. Equation 7-6 is shown in Figure 7-8.

Open Tube: For this filter, shown in Figure 7-7*b*, the transmission loss is

$$TL = 10\log_{10}\left[1 + 0.25\cot^2 kL'\right] \qquad dB \qquad (7-8)$$

where L' is given by (7-7). At low frequencies ($L' \ll \lambda$) the behavior of the tube is similar to that of the orifice. Whenever $kL' = n\pi$ ($L'/\lambda = n/2$) where n is an integer, the TL is a maximum. Whenever $kL' = (2n-1)\pi/2$, the TL is a minimum. Equation 7-8 is plotted in Figure 7-9.

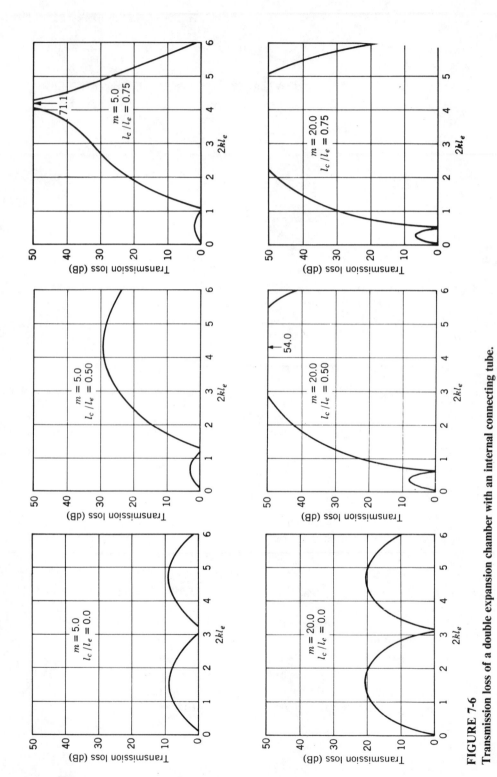

FIGURE 7-6
Transmission loss of a double expansion chamber with an internal connecting tube.

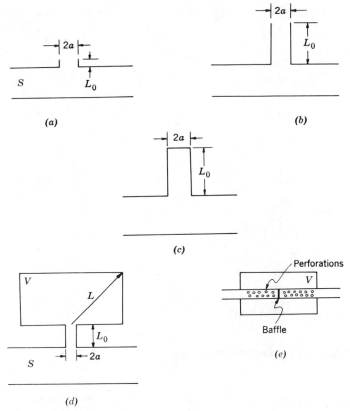

FIGURE 7-7
Schematic diagrams of various side-branch filters: (a) orifice; (b) open tube; (c) closed tube; (d) resonator; (e) single chamber and perforated tube.

Closed Tube: For this filter, shown in Figure 7-7c, the transmission loss is

$$TL = 10\log_{10}[1 + 0.25\tan^2 kL']   \text{dB} \tag{7-9}$$

where L' is again given by (7-7). Whenever $kL' = n\pi$, the TL is a minimum and when $kL' = (2n - 1)\pi/2$, it is a maximum. As $kL' \to 0$, TL $\to 0$ so that the closed tube branch acts as a low-pass filter at low frequencies, whereas the open-tube branch acts as a high pass filter at low frequencies. Equation 7-9 is plotted in Figure 7-10.

Resonators: For this filter, shown in Figure 7-7d, the transmission loss is

$$TL = 10\log_{10}\left[1 + \mu^2\beta^2(f/f_r - f_r/f)^{-2}\right]   \text{dB} \tag{7-10}$$

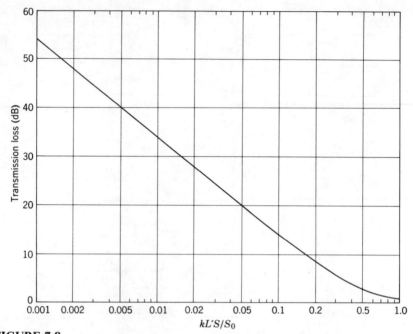

FIGURE 7-8
Transmission loss of an orifice side-branch filter.

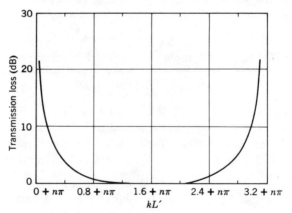

FIGURE 7-9
Transmission loss of an open tube side-branch filter ($n = 0, 1, 2, \ldots$).

224

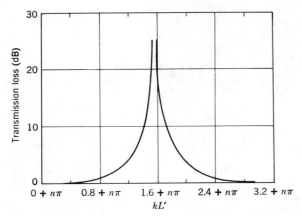

FIGURE 7-10
Transmission loss of a closed tube side-branch filter ($n = 0, 1, 2, \ldots$).

where

$$f_r = \frac{c}{2\pi} \sqrt{\frac{c_0}{V}} \qquad \text{Hz}$$

$$\mu = \frac{\sqrt{c_0 V}}{2S}$$

$$c_0 = \frac{S_0}{L'} = \frac{\pi a^2}{L'}$$

and V is the volume of the resonator and L' is given by (7-7). The quantity β is equal to unity if air flow is neglected. If flow is considered, then[4]

$$\beta = 2.86 \bar{p} \frac{1 - \bar{p}}{\overline{M}} \qquad (7\text{-}11)$$

where $\overline{M}$ is the mean flow Mach number in the pipe and $\bar{p} = \Delta p / p_a$ is the pressure drop fraction; p_a is the ambient pressure and $\Delta p = p_o - p_a$ where p_o is the mean pressure inside the muffler in the same units as p_a. Equation 7-10 is plotted in Figure 7-11. Equation 7-10 is valid only when $L + L_0 < \lambda/8$. For the case of n orifices of length L_0 and area S_0, terminating in the same volume, c_0 is given by

$$c_0 = \frac{n S_0}{L_0 + 0.8 \sqrt{S_0}}$$

Single Chamber and Perforated Tube: The following formula for the

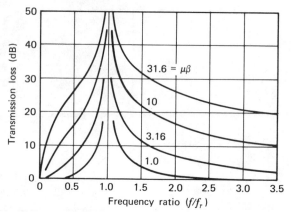

FIGURE 7-11
Transmission loss of a single chamber resonator side-branch filter.

transmission loss of this type of muffler, shown in Figure 7-7e, has been obtained[4] in a form that is related to the characteristics of a diesel or gasoline engine:

$$TL = 10 \log_{10} \left[1 + \left(\frac{8\pi\bar{p}}{1+3\bar{p}} \frac{V}{V_c} \right)^2 \right] \qquad dB \qquad (7\text{-}12)$$

In (7-12) V_c is the corrected displacement (volume) of one cylinder of the engine given by $V_c = 2.5\, r_c V_e / N$ where N is the number of cylinders of the engine, V_e is the engine displacement, $\bar{p}$ is defined as for (7-11), and r_c is the turbocharger pressure ratio, which equals unity for a normally aspirated engine. A plot of (7-12) is shown in Figure 7-12, which clearly illustrates how the filter affects the pressure drop (i.e., adds a back pressure) to the exhaust system.

Right Angle Bend in Tube with Square Cross Section

The transmission loss of an elbow duct with a square cross-section of width d is

$$TL = 10 \log_{10} \left\{ \frac{k^2 d^2}{4} \left[\left(1 + \frac{1}{k^2 d^2} - \cot^2 kd \right)^2 + 4 \cot^2 kd \right] \right\} \qquad dB \quad (7\text{-}13)$$

which is valid for $kd \leqslant 2.44\pi$. Equation 7-13 is plotted in Figure 7-13.

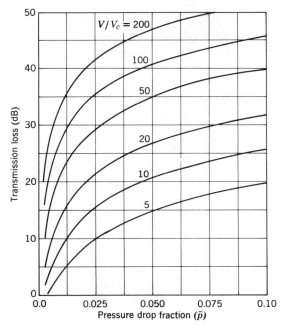

FIGURE 7-12
Transmission loss of a side-branch filter having a single chamber with a perforated tube.

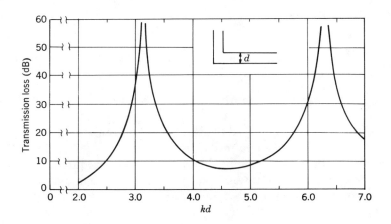

FIGURE 7-13
Transmission loss of an elbow duct with a square cross-section.

Radiated Sound from an Engine Exhaust Pipe

The above results were obtained using classical acoustic plane wave theory. The actual performance of silencers has often been inferior to that expected. It has been shown[5] that the sound-pressure level of the sound radiated from a tail pipe will be considerably underestimated unless the mean flow of the gas is also included in the analysis. If R is the reflection coefficient, defined as the ratio of the magnitude of the sound pressure reflected from the outlet of the tail pipe to the sound pressure incident, then the rms amplitude of a spherical pressure wave $P_{o_{rms}}^2$, at a radius r from the exhaust outlet is given by

$$P_{o_{rms}}^2 = \left(\frac{\rho_o c_o}{\rho_i c_i} \right) \frac{a^2}{4r^2} P_{i_{rms}}^2 \left[(1+M)^2 - R^2(1-M)^2 \right] \qquad (7\text{-}14)$$

where $P_{i_{rms}}^2$ is the amplitude of the incident pressure wave at the exhaust outlet and M is the mach number of the exhaust flow. The subscripts i and o refer to the inside and the outside of the pipe, respectively. If the mean flow is neglected, $M=0$ and (7-14) reduces to that predicted by classical theory.

Transmission Loss of Dissipative Filters

Attenuation in Lined Ducts*

The attenuation of sound in a lined duct may be computed from the following empirical formula:

$$L_A = 12.6 \frac{D_0}{S} \alpha_{SAB}^{1.4} \qquad \text{dB/m} \qquad (7\text{-}15)$$

where L_A is the attenuation in dB/m, D_0 is the perimeter of the duct liner (m), S is the area of the duct (m²), and α_{SAB} is the Sabine absorption coefficient for the liner material. This formula is accurate within 10% for ducts having cross-sectional dimensions in the ratio of $1:1$ to $2:1$, for absorption coefficients between 0.20 and 0.40, and frequencies between 250 and 2000 Hz. The absorption coefficient appropriate in this calculation should be obtained from reverberation measurements where the absorptive material is backed with sheet metal and supported on 25 by 76 mm furring strips, 610 mm on centers.

*Reference 1, pp. 27-9 to 27-11.

Examination of (7-15) shows that if the area, S, and the absorption coefficeint are held constant, the total attenuation will be proportional to the product of the duct perimeter and length of lining. Consequently, the use of splitters or "egg-crate" type separaters to increase the D_0/S ratio results in a more compact attenuator. In computing the attenuation from a splitter-type liner the absorption coefficient for a thickness of material equal to one-half the thickness of the splitter is used. This consideration is most important at the low frequencies, where the absorption is strongly dependent on the thickness of the absorber and where all the energy is carried by sound waves moving parallel to the duct walls. Therefore, to be equally effective, splitters should be twice as thick as the material used along the duct boundaries. These ideas are illustrated in Figure 7-14.

The results for a lined duct with a 180° bend are shown in Figure 7-15. It should be noted that Figure 7-15 is only valid where the lining is applied to the sides which undergo the 90° bend out-of-plane.

Duct Silencers

In air handling equipment the ability of lined ducts to attenuate sound is very important. In order to obtain an idea of their capabilities typical curves[6,7] of various geometries are presented in Figures 7-16 through 7-19. Figure 7-16 shows a design often used for fan or turbine intakes. The three curves are for lengths of 1.2, 2.4, and 3.7 m. The high frequencies can be greatly attenuated if the absorbing materials are staggered as shown in Figure 7-17. Notice that the 1.2 m section of Figure 7-17 is equal to the 1.2 m section shown in Figure 7-16. In both these figures the open (free) area is equal to 50% of the total cross-sectional area of the duct. To improve low frequency performance at the expense of high-frequency performance the thickness of the lining is increased, thereby reducing the open area of the duct. An example of this is shown in Figure 7-18 where the open area is 33%. For good low and high frequency attenuation a combination of the configuration shown in Figure 7-18 with that shown in Figure 7-17 will suffice. Typical results for this configuration is shown in Figure 7-19.

Lined Plenum Chambers

The geometry of a single plenum chamber is shown in Figure 7-20. The theoretically derived transmission loss is [see (6-14) and (6-19)]

$$\text{TL} = -10\log_{10}\left[S\left(\frac{H}{2\pi q^3} + \frac{1 - \alpha_{\text{ST}}}{S_L \alpha_{\text{ST}}} \right) \right] \quad \text{dB} \quad (7\text{-}16)$$

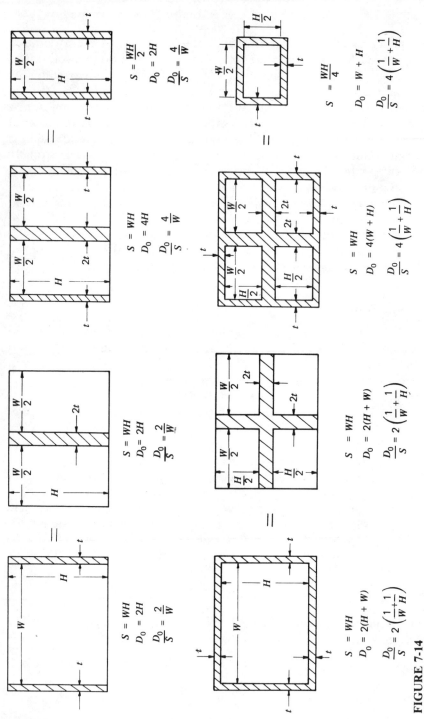

FIGURE 7-14
Duct-liner configurations having equal attenuation.

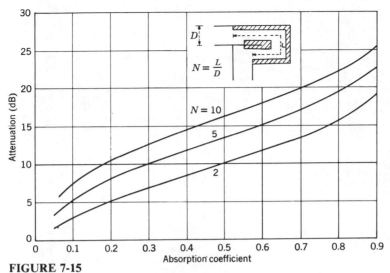

FIGURE 7-15
Attenuation of a lined duct with a 180° bend.

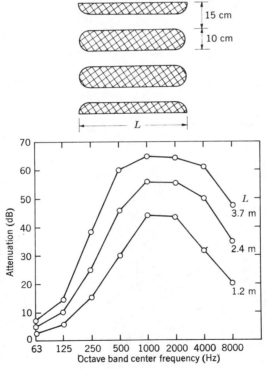

FIGURE 7-16
Attenuation in ducts containing thin parallel baffles. (From References 6 and 7.)

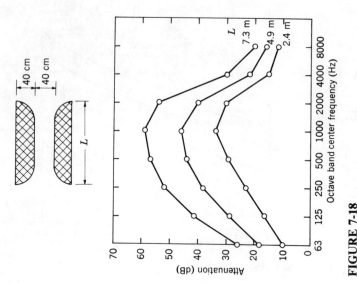

FIGURE 7-18
Attenuation in ducts containing thick parallel baffles.
(From References 6 and 7.)

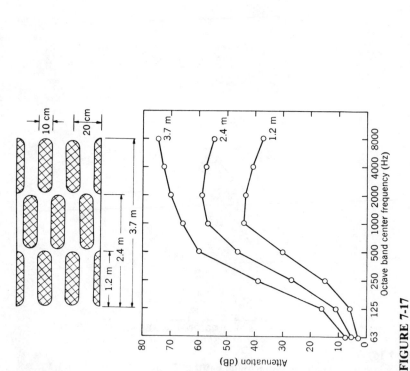

FIGURE 7-17
Attenuation in ducts containing thin, staggered parallel baffles.
(From Refernces 6 and 7.)

232

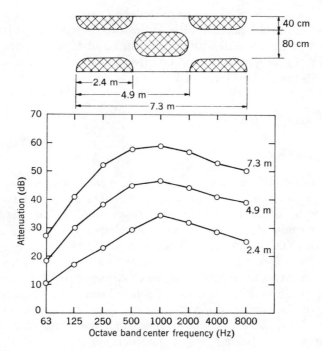

FIGURE 7-19
Attenuation in ducts containing thick staggered parallel baffles. (From References 6 and 7.)

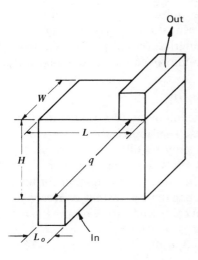

FIGURE 7-20
Single plenum chamber.

233

where S is the area of the plenum outlet, S_L is the area of the acoustic lining, and α_{ST} is the random incident absorption coefficient [recall (6-3)]. Equation 7-16 agrees well with measurements at high frequencies and for values of L_o/L not too large. At low frequencies (7-16) yields values that are 5–10 dB lower than measured values.

7.2   BARRIERS

Introduction

Noise control by barriers is a common means of obtaining a modest reduction in the overall sound-pressure level. Traffic noise from highways and railroads, other outdoor noises from construction machinery or stationary installations, such as large transformers or plants, and indoor noise in open plan offices and schools can be shielded by a barrier, which intercepts the line-of-sight from the source to a receiver. The noise at the receiver is reduced to that portion which arrives via diffraction over the barrier top or around its ends, via reflection from other buildings, and via scattering and refraction in the atmosphere. In addition to the measurable acoustical effect, there is much evidence that the visual shielding of the noise source by a barrier has a considerable psychological effect. In outdoor noise control, a row of trees, which is known to have a marginal influence on sound propagation, ranks among the most frequently desired protective measures.

It is the purpose of this section to present some of the design relationships that have evolved in recent years. The information contained in this section can be supplemented by referral to the references in Kurze's[8] thorough review of the literature of the noise reducing properties of barriers. In addition, two theoretical investigations[9,10] concerning wide barriers and the placement of absorbing materials on barrier surfaces have recently been published which may lead to improved barrier designs.

Very Long Barriers Outdoors

Nonporous walls of sufficient mass (minimum of 10 kg/m²), if interposed between source and receiver, can result in appreciable noise reduction, because sound can reach the receiver only by diffraction around the boundaries of the obstacle. The sound interacts with the barrier in three ways: (1) reflection back from its surface; (2) direct transmission through it; and (3) diffraction over its top. The propagation path that concerns us in this section is that of diffraction. However, before proceeding with the results obtained from diffraction theory it is worthwhile to examine the second propagation path, namely, transmission through the barrier. (This topic is discussed in detail in Section 7.4.) To illustrate its relative importance consider the following: let the attenuation of the source level due to the diffraction of the sound over the top of the barrier be ΔL_D and that caused by the transmission loss through the barrier ΔL_T. Then from (1-25) the original sound-pressure level has been attenuated by the amount

$$\Delta L = \Delta L_D - 10 \log_{10}[1 + 10^{-(\Delta L_T - \Delta L_D)/10}] \quad \text{dB}$$

The larger ΔL is, the greater the attenuation; therefore, we want the term containing the logarithm to be as small as possible. It is easily seen that when $\Delta L_T - \Delta L_D \geqslant 9$ dB, $\Delta L \geqslant \Delta L_D - 0.5$. In other words, the transmission loss properties of the barriers need be only 9 dB greater than the desired barrier attenuation ΔL_D in order that it influence the barrier's attenuating ability by less than 0.5 dB. For example, if the desired barrier attenuation is 15 dB ($= \Delta L_D$), the transmission loss properties of the barrier should be greater than 24 dB ($= \Delta L_T$).

Maekawa[11] has investigated both theoretically and experimentally the reduction of sound of a point source (recall p. 4) by an infinitely long barrier. These results have been extended[12] to consider the attenuation of sound of an infinitely long incoherent line source and their validity has been investigated experimentally.[13] All these investigations ignore the effects of wind and the possible attenuation of the ground cover. The results are summarized in Figure 7-21 where it can be seen that the

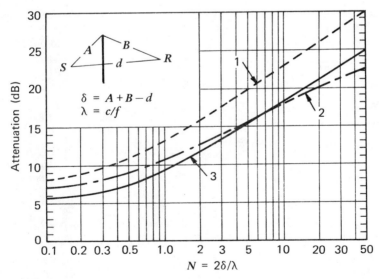

FIGURE 7-21
Sound attenuation of a barrier: (1) Maekawa's[11] result for a point source; (2) Kurze and Anderson's[12] result calculated for a line source; (3) experimental results of Koyasu and Yamashita.[13] (Reproduced from Applied Acoustics, Volume 6, No. 3 by courtesy of the publisher and authors.)

expected attenuation of a barrier for a line source (which is a reasonably good representation of freely flowing heavy traffic) is 3–5 dB less than that expected for a single point source.

Scholes et al.[14,15] have used Maekawa's results to convert typical octave band L_{10} traffic noise levels into A-weighted-sound levels. These theoretical results were experimentally verified for various realistic conditions including a wind of 2–3 m/sec from source (roadway) to receiver. Their results are summarized in Figures 7-22 and 7-23. The elevated situation refers to the case where either the road or the reception point, or both, are

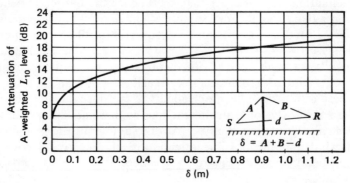

FIGURE 7-22
Attenuation of A-weighted sound level for elevated positions. (Can also be used in flat situations with a wind component of 2–3 m/sec from road to receiver.) (Reproduced from Applied Acoustics, Volume 5, No. 3 by courtesy of the publisher and authors.)

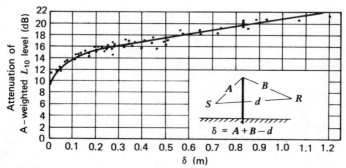

FIGURE 7-23
Attenuation of A-weighted sound level for flat situations and no wind. (Points are calculated from experimental data.) (Reproduced from Applied Acoustics, Volume 5, No. 3 by courtesy of the publisher and authors.)

elevated above the general ground level so that the mean height of the propagation path without a barrier is 6 m or more above the ground. In these situations, and for reception positions within 120 m, the influence of the ground will be negligible for practical purposes and the effect of the wind component will be reduced. The flat situation covers those cases in which both the road and the reception point are close enough to the general ground level for the mean height of the propagation path, top of the barrier to receiver, to be less than 6 m. In these situations, both ground and wind effects can be expected to influence barrier performance.

It should be mentioned that when the barrier is effectively the side of a depressed roadway, where the depth of the highway depression is proportional to the height of the barrier, care must be exercised in the angle of slope of the cut walls. Figure 7-24a illustrates the point. Figure 7-24b shows how the reflection off the far wall can be eliminated. However, it should be noted that the effective barrier height is also reduced for the direct sounds. The effective barrier height can be increased again by erecting a low barrier wall on top of the embankment.

Practical information concerning the actual details of barrier construction, their estimated costs, and the expected or measured attenuation of the A-weighted noise levels has been compiled by the Federal Highway Administration.[16] Also, two detailed studies[17,18] concerning the effects of trees, shrubs, and atmospheric conditions on the attenuation of sound are available.

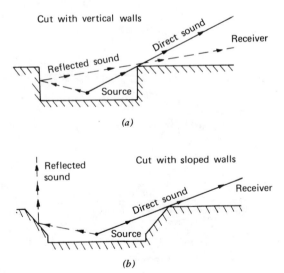

(a)

(b)

FIGURE 7-24
Effect of sloping the walls of a depressed roadway on the sound reaching the receiver.

Prediction of Barrier Attenuation of Traffic Noise Using a Nomograph

A nomograph procedure has been developed[19] whereby the attenuation of A-weighted traffic noise levels by either a finite or infinite barrier of uniform height that is parallel to an infinite, straight roadway can be estimated. The nomograph shown in Figure 7-25 can be used for level, elevated, or depressed roadways, as depicted in Figure 7-26. It should be noted that in general the length of the line-of-sight (L/S) is *not* equal to the horizontal distance between source and receiver and that the "break"

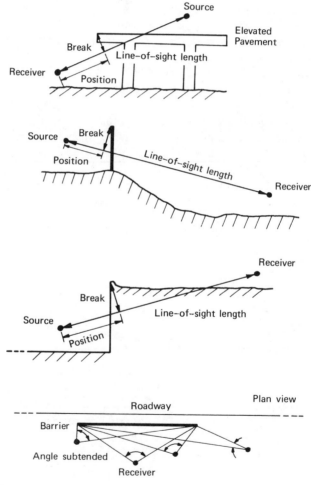

FIGURE 7-26
Barrier parameters for various roadway configurations.

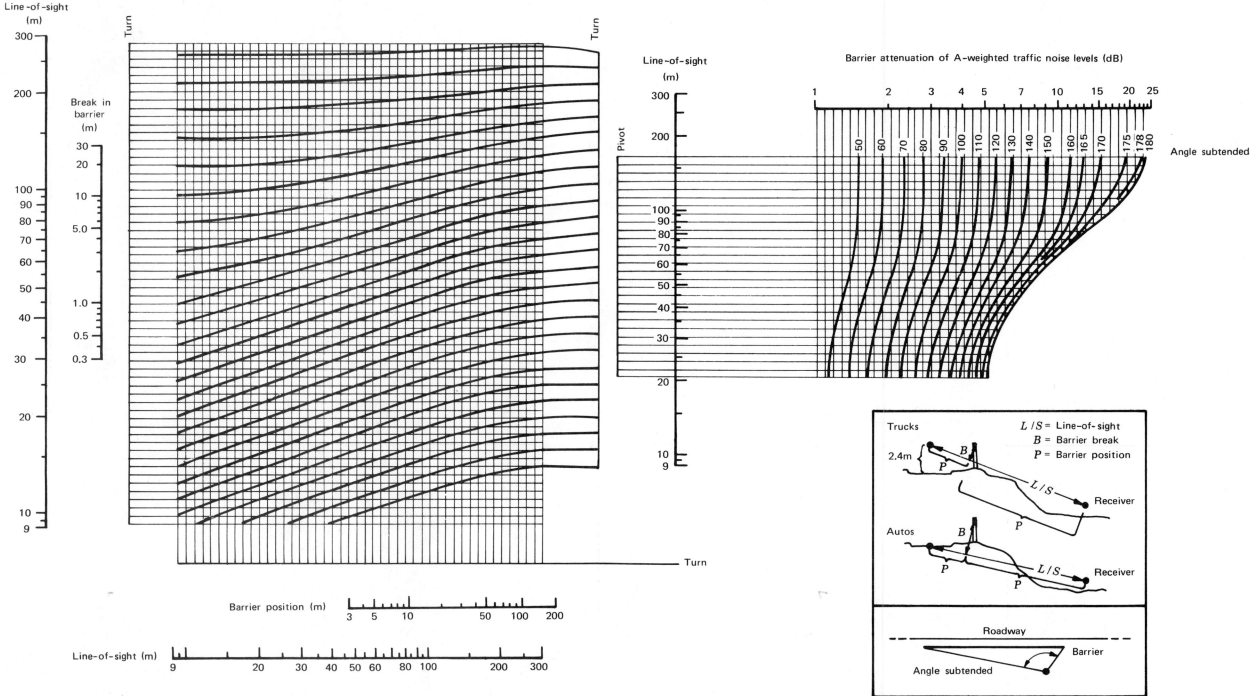

FIGURE 7-25 Nomograph to estimate the attenuation by a barrier of the A-weighted sound level of highway traffic noise. (From Reference 19.)

in the L/S is *not* necessarily equal to the height of the barrier. It should be noticed that the position for the center of the source for trucks is placed 2.44 m *above* the road surface whereas the center for autos is at the roadway surface. The use of the nomograph is illustrated in Example 7-2.

Example 7-1.   A machine, which can be considered an omnidirectional source, is to be placed 1 m off the ground and 4 m behind a very long barrier. The machine emanates a narrow band of noise centered about 500 Hz. What must be the minimum height of the barrier in order that the noise from the machine be reduced 13 dB at a point 1.5 m off the ground and 8 m from the other side of the barrier.

Solution: The wavelength of the sound in air at standard conditions at 500 Hz is $\lambda = c/f = 344/500 = 0.688$ m. From curve 1 of Figure 7-21 $N = 0.95$. From the position of the source and receiver it is a relatively easy matter to show that $A = [(H-1)^2 + 4^2]^{1/2}$, $B = [(H-1.5)^2 + 8^2]^{1/2}$, and $d = [12^2 + 0.5^2]^{1/2} = 12.01$ m, where A, B, and d are defined in Figure 7-21 and H is the height of the barrier. Thus

$$N = \frac{2}{0.688}\left[\sqrt{(H-1)^2 + 4^2} + \sqrt{(H-1.5)^2 + 8^2} - 12.01\right] = 0.95$$

or

$$12.34 = \sqrt{(H-1)^2 + 4^2} + \sqrt{(H-1.5)^2 + 8^2}$$

Trial and error yields $H = 2.5$ m.

Example 7-2.   Consider a single lane of automobiles shielded by an earth berm, as shown in Figure 7-27. The berm subtends an angle of 180° at the receiver location. Using the nomograph of Figure 7-25 determine the barrier attenuation.

Solution: The line-of-sight is 38.1 m and the break is 3 m. The smaller of the "positions" is 15.2 m. Using the nomograph redrawn in Figure 7-28 it is found that the A-weighted level is attenuated 12.5 dB.

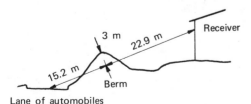

FIGURE 7-27
Barrier geometry for Example 7-2.

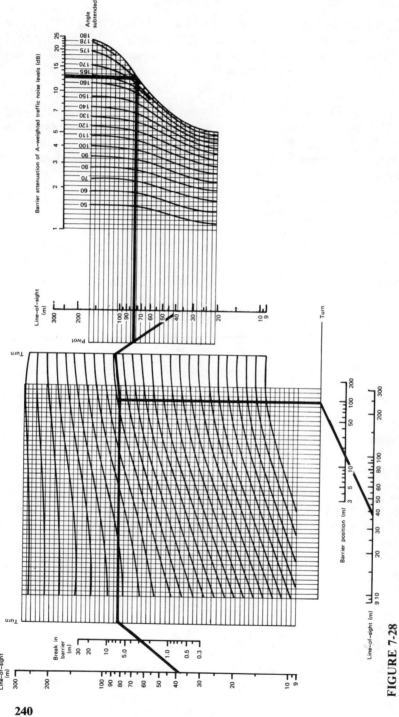

FIGURE 7-28
Illustration of nomograph usage for barrier configuration of Figure 7-27.

240

Example 7-3. Repeat Example 7-2 using Figure 7-23.

Solution: From Figure 7-27 it is easily shown that the A, B, d of Figure 7-23 are: $A = 15.5$ m, $B = 23.1$ m, and $d = 38.1$ m. Therefore, the path difference $A + B - d = 0.5$ m. From Figure 7-23 the infinitely long barrier will yield an attenuation of the A-weighted level of approximately 17.5 dB. It is seen that the two methods differ by 5 dB with the nomograph procedure being more conservative in this particular example. It should be noted, however, that Figure 7-23 is based on experimental data of actual British traffic conditions whereas the nomograph is based in part on American traffic conditions and in part on theoretically derived values.

7.3 VIBRATION ISOLATION

Basic Vibration Theory[20-22]

Vibratory systems are comprised of means for storing potential energy, (the spring), means for storing kinetic energy (mass or inertia), and means by which energy is gradually lost (the damper). The vibration of a system involves the alternating transfer of energy between its potential and kinetic forms. In a damped system, some energy is dissipated at each cycle of vibration and must be replaced from an external source if a steady vibration is to be maintained. Although a single physical structure may store both kinetic and potential energy, and may dissipate energy, this section considers only lumped parameter systems comprised of ideal springs, masses, and dampers, wherein each element has only a single function. In translation motion, displacements are defined as linear distances; in rotational motion, displacements are defined as angular motions.

The simplest possible vibratory system is shown in Figure 7-29. It consists of a mass m attached by means of a spring k and a viscous damper c to an immovable support. In the linear spring the change in length of the spring is proportional to the force acting along its length. This idealized spring is considered to have no mass.

If the spring k in Figure 7-29 were not attached to a rigid support, but instead to one which had an equivalent stiffness, k_1, k would be replaced by a new stiffness value, k_0, given by

$$k_0 = \frac{kk_1}{k_1 + k} \qquad (7\text{-}17)$$

In (7-17) it is seen that when $k_1 \gg k$, $k_0 \approx k$; that is, the stiffness represented by k_1 is relatively rigid compared to k, which is the original case. When

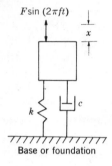

$F\sin(2\pi ft)$

x

k

c

FIGURE 7-29

Single degree-of-freedom system with a viscous damper.

Base or foundation

$k_1 \ll k$, $k_0 \approx k_1$; that is, the original spring of the lumped system is relatively rigid compared to that having a stiffness of k_1.

If W is the weight of the rigid body forming the mass of the system m, then $W = mg$ where g is the acceleration of gravity. The static deflection δ_{ST} due to W is

$$\delta_{ST} = \frac{W}{k} = \frac{mg}{k} \qquad (7\text{-}18)$$

Now consider the case wherein the mass m in Figure 7-29 is excited with a force that varies harmonically with time, that is, $F\sin 2\pi ft$. The magnitude of the displacement of the mass, denoted x, at any frequency f is

$$x = \frac{F}{kZ_0} \qquad (7\text{-}19)$$

where

$$Z_0 = \left\{ \left[1 - \left(\frac{f}{f_n} \right)^2 \right]^2 + \left[2\xi \frac{f}{f_n} \right]^2 \right\}^{1/2} \qquad (7\text{-}20)$$

$$f_n = \frac{1}{2\pi}\sqrt{\frac{k}{m}} = \frac{1}{2\pi}\sqrt{\frac{g}{\delta_{ST}}} \quad \text{Hz} \qquad (7\text{-}21)$$

and

$$\xi = \frac{c}{c_c} = \frac{c}{(4\pi f_n m)} \qquad (7\text{-}22)$$

The quantity f_n is called the natural frequency of the system, which is seen

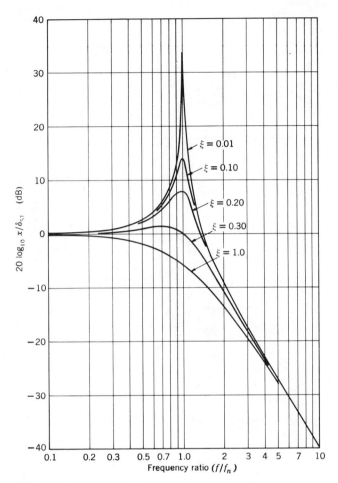

FIGURE 7-30
Relative displacement response of a single degree-of-freedom system as a function of the ratio of the forcing frequency to its undamped natural frequency.

to be only a function of the static deflection of the spring, and ξ is called the fraction of the critical damping c_c. Equation (7-19) is shown in Figure 7-30. (Equation 7-21 is shown in Figure 7-39.)

Force Transmission (Transmissibility)

Let us examine the force transmitted to the foundation when the mass in Figure 7-29 is excited by a harmonically varying force $F \sin 2\pi ft$. If the magnitude of the transmitted force is F_T, the transmissibility, T, can be shown[22] to be

$$T = \frac{|F_T|}{F} = \frac{\left[1 + \left(2\xi\frac{f}{f_n}\right)^2\right]^{1/2}}{Z_0} \qquad (7\text{-}23)$$

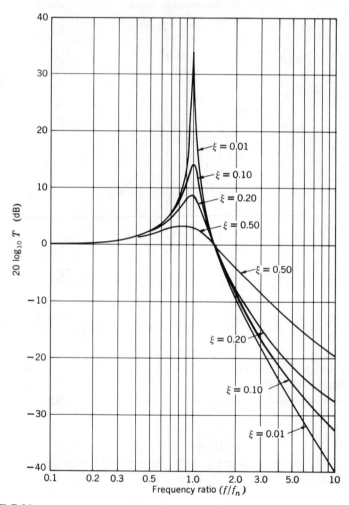

FIGURE 7-31
Transmissibility of a single degree-of-freedom system as a function of the ratio of the forcing frequency to its undamped natural frequency. (Force and displacement transmissibility are numerically identical.)

where Z_0 is given by (7-20). Equation (7-23) is shown in Figure 7-31. If one were to excite the base (foundation) of the spring-mass-damper system with a displacement that varied harmonically with time instead of its mass, as indicated above, we would find the same relation (7-23) for the transmissibility. Hence Figure 7-31 portrays the displacement transmissibility as well as the force transmissibility.

From Figure 7-31 it is seen that regardless of the damping, the transmissibility is always greater than unity for $f/f_n < \sqrt{2}$. When $f/f_n > \sqrt{2}$, $T < 1$. It is also noticed that for $f/f_n > \sqrt{2}$, T is the smallest when ξ is the smallest. Thus for vibration isolation, one wants to be farthest away from the natural frequency of the system $(f \gg f_n)$; the isolation system should also have as little viscous damping as possible $(\xi \to 0)$. It is also seen from Figure 7-31 that for $\xi \leqslant 0.05$ the transmissibility is one-tenth (-20 dB) of that at $f = 0$ when the forcing frequency is 3.22 times greater than the natural frequency of the system. The transmissibility decreases to one-hundredth (-40 dB) of that at $f = 0$ when the forcing frequency is slightly more than 10 times greater than the natural frequency of the system.

Several-Degree-of-Freedom-System

The single degree of freedom system shown in Figure 7-29 is adequate for illustrating the fundamental principles of vibration, but is a great over-simplification insofar as many practical applications are concerned. Consider the asymmetrical body shown in Figure 7-32, which is supported by springs at each corner of its base. When the resilient supports are located unsymmetrically with respect to the center of gravity of the body, certain translatory and rotational modes of vibration may become coupled. For example, in Figure 7-32 the vibration in the vertical translatory mode becomes coupled to vibration in a rotational mode in which the mounted body vibrates with respect to an axis that doesn't coincide with the center of gravity of the body.

A body on resilient supports may vibrate in one natural mode simultaneously with, but independently of, its vibration in other natural modes if the respective modes are decoupled. If the modes are coupled, vibration in one of the coupled modes cannot occur independently of vibration in another of the modes. Coupling depends on the stiffness and location of the resilient supports and on the mass distribution of the supported body. To determine whether translatory vibration in a particular direction is decoupled, apply a steady force to the body through its center of gravity and in the specified direction. If the body moves in the direction of the force without rotation, translatory vibration in the direction of the force is decoupled from vibration in other modes.

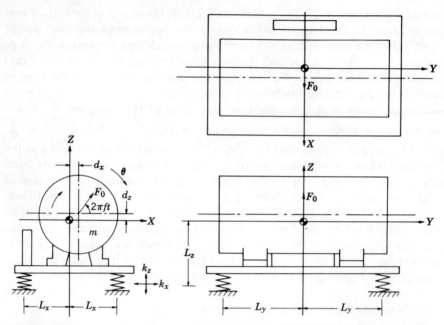

FIGURE 7-32
Rigid body of mass m mounted on flexible supports located at its four corners and having two planes of symmetry with respect to the location of the center-of-gravity and the spring placements.

Now consider the system shown in Figure 7-32 subjected to the following forces in the ZY-plane: $F_x = F_0 \cos 2\pi ft$ and $F_y = F_0 \cos 2\pi ft$. This represents a force caused by an unbalanced rotating element of a machine. The vertical displacement is the Z-direction is uncoupled, and is given by (7-19) with $\xi = 0$. The transmissibility in this direction is given by (7-23) with $\xi = 0$. The horizontal displacement amplitude x_c of the center-of-gravity in the X-direction and the rotational displacement amplitude θ_c about the Y-axis are given by*

$$x_c = \left(\frac{F_0}{4k_x} \right) \left(\frac{K}{D_0} \right) A_1$$

$$\theta_c = \left(\frac{F_0}{4k_x r_y} \right) \left(\frac{K}{D_0} \right) A_2$$

(7-24)

*Reference 20, p. 3-35.

where

$$A_1 = \left\{ \left[K\left(\frac{L_z}{r_y}\right)\left(\frac{L_z}{r_y} - \frac{d_z}{r_y}\right) + 1 - \hat{f}^2 \right]^2 + \left(K\frac{L_z}{r_y}\frac{d_x}{r_y}\right)^2 \right\}^{1/2}$$

$$A_2 = \left\{ \left[K\left(\frac{L_z}{r_y} - \frac{d_z}{r_y}\right) + \frac{d_z}{r_y}\hat{f}^2 \right]^2 + \left[\frac{d_x}{r_y}(K - \hat{f}^2) \right]^2 \right\}^{1/2}$$

$$D_0 = \hat{f}^4 - \left\{ K\left[1 + \left(\frac{L_z}{r_y}\right)^2 \right] + 1 \right\}\hat{f}^2 + K \qquad (7\text{-}25)$$

$$K = \frac{k_x}{k_z}\left(\frac{r_y}{L_x}\right)^2$$

$$\hat{f} = \frac{f}{f_z}\frac{r_y}{L_x} \qquad f_z = \frac{1}{2\pi}\sqrt{\frac{4k_z}{m}}$$

and r_y is the radius of gyration of the mass m about the Y-axis. The ratio k_x/k_z has been determined* for steel springs and varies between approximately 0.25 and 2.0 depending upon the spring's dimensions and its static deflection.

The natural frequencies of the coupled system are found by setting $D_0 = 0$; thus

$$\hat{f}_{1,2} = \frac{1}{\sqrt{2}}\left\{ 1 + K\left(1 + \frac{L_z^2}{r_y^2}\right) \pm \left[\left\{ 1 + K\left(1 + \frac{L_z^2}{r_y^2}\right) \right\}^2 - 4K \right]^{1/2} \right\}^{1/2} \qquad (7\text{-}26)$$

Equation 7-26 is shown in Figure 7-33. Two numerically different values of the dimensionless frequency ratio $\hat{f}$ correspond to the two discrete coupled modes of vibration. The two straight lines in Figure 7-33 for $L_z/r_y = 0$ represent natural frequencies in decoupled modes of vibration. When $L_z = 0$, the elastic supports lie in a horizontal plane passing through the center of gravity of the support body. The horizontal line at a value of unity on the ordinate scale represents the natural frequency of the uncoupled rotational mode. The inclined straight line for the value $L_z/r_y = 0$ represents the natural frequency of the system in horizontal translation.

*Reference 20, pp. 34-4 to 34-22.

When $L_z/r_y = 0$ the two uncoupled natural frequencies of the system assume their lowest values. Therefore, in a practical situation it is advisable to shift the center-of-gravity to a position such that $L_z = 0$. This is shown in Figure 7-34 where the center-of-gravity was also shifted simultaneously laterally so that $d_x = 0$. Examination of (7-25) shows that when both $L_x = d_x = 0$ the magnitude of A_1 and A_2 is reduced, thereby reducing the displacement x_c and rotation θ_c.

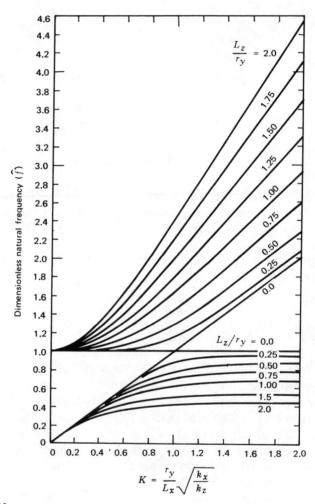

FIGURE 7-33
Dimensionless natural frequencies $\hat{f}$ of the coupled vibrations of the configuration of Figure 7-32 as a function of the parameter K.

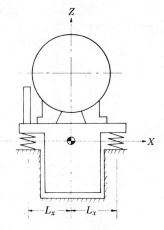

FIGURE 7-34
Modification of the system shown in Figure 7-32 so that $L_z = 0$. This is a system with three planes of symmetry with respect to the location of the center-of-gravity to the spring placements.

Dynamic Absorbers

Consider the system shown in Figure 7-35 consisting of a primary system of k and m to which an auxiliary system k_a, c_a, and m_a have been added. The system in Figure 7-35b is equivalent to the system shown in Figure 7-35a since $F = kU$. Solving for the magnitude of the amplitude of motion, x_0, of the primary system shown in Figure 7-35a yields

$$\frac{x_0}{\delta_{ST}} = \left\{ \frac{(\alpha^2 - \beta^2)^2 + (2\xi\alpha\beta)^2}{\left[(\alpha^2 - \beta^2)(1 - \beta^2) - \alpha^2\beta^2\mu\right]^2 + (2\xi\alpha\beta)^2(1 - \beta^2 - \beta^2\mu)^2} \right\}^{1/2} \tag{7-27}$$

where

$$\mu = \frac{m_a}{m} \qquad\qquad \alpha = \frac{f_a}{f_n}$$

$$f_n = \frac{1}{2\pi}\sqrt{\frac{k}{m}} \qquad \beta = \frac{f}{f_n} \tag{7-28}$$

$$f_a = \frac{1}{2\pi}\sqrt{\frac{k_a}{m_a}} \qquad \xi = \frac{c}{c_c} = \frac{c}{(4\pi m_a f_a)}$$

and $\delta_{ST} = F/k$. When $\xi = 0$, (7-27) reduces to

$$\frac{x_0}{\delta_{ST}} = \frac{\alpha^2 - \beta^2}{(\alpha^2 - \beta^2)(1 - \beta^2) - \mu\alpha^2\beta^2} \tag{7-29}$$

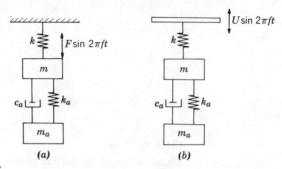

FIGURE 7-35
Coupling of an auxiliary spring-mass-dashpot system to a primary spring-mass system.

Thus when $\xi = 0$, $x_0 = 0$ when $\beta = \alpha$ ($f = f_a$). The vibration of the primary system is thereby eliminated entirely when the auxiliary system is undamped and is tuned to the forcing frequency.

Although the concept of tuning a dynamic absorber appears simple, practical considerations make it difficult to tune any such system exactly. When the auxiliary mass is small relative to the primary mass, m, its effectiveness depends on accurate tuning. If the tuning is incorrect, the addition of the auxiliary mass may bring the composite system (primary and auxiliary systems) into resonance with the exciting force.

The natural frequencies of the composite system of Figure 7-35 are determined from

$$\frac{f_n}{f_a} = \frac{1}{\alpha\sqrt{2}}\left\{\alpha^2(1+\mu)+1 \pm \sqrt{-4\alpha^2+\left[\alpha^2(1+\mu)+1\right]^2}\right\}^{1/2} \quad (7\text{-}30)$$

Because of the introduction of the auxiliary mass the composite system now has two natural frequencies instead of one, even though m_a was added to eliminate the amplitude at $f = f_a$. Since the absorber is nominally tuned to this frequency of excitation, the root from (7-30) that is closer to the forcing frequency is of interest. Hence f_n/f_a is a measure of the sensitivity of the tuning required to avoid one of these two resonances. Equation 7-30 is equal to (7-26) if the substitutions $K = \alpha^2$, $L_z^2/r_y^2 = \mu$, and $\hat{f} = \alpha f_n/f_a$ are made. Therefore, with these substitutions (7-30) is also given by Figure 7-33.

Dynamic absorbers are most generally used when the primary system without the absorber is nearly in resonance with the excitation. If the natural frequency of the primary system is less than forcing frequency, it is preferable to tune the dynamic absorber to a frequency slightly lower than the forcing frequency to avoid the resonance that lies above the natural frequency of the primary system. Likewise, if the natural frequency of the primary system is above the forcing frequency, it is well to tune the damper to a frequency slightly greater than the forcing frequency.

Where the natural frequency of the composite system is nearly equal to the tuned frequency of the absorber, the amplitude of motion of the primary mass to resonance is much smaller than that of the absorber. Consequently, the motion of the primary mass does not become large even at resonance; but the motion of the absorber, unless limited by damping, may become very large.

Auxiliary Mass Dampers

In a multiple degree of freedom system, the introduction of an auxiliary mass system tends to lower those original natural frequencies of the primary system that are below the tuned frequency of the auxiliary system. The original natural frequencies that are higher than the tuned frequency of the auxiliary system are raised by adding the auxiliary mass system. A new natural mode of vibration corresponding to the vibration of the auxiliary mass system against the primary system is injected between the displaced initial natural frequencies of the primary system. Those frequencies of the primary system that are closest to the tuned frequency are most strongly influenced by the auxiliary mass system. The addition of damping in the auxiliary mass system can be effective in reducing the amplitude of motion of the primary system at the natural frequencies. For this reason auxiliary mass dampers are used quite commonly to reduce overall vibration stresses and amplitudes.

For the auxiliary mass damper to be most effective in limiting the value of x_0/δ_{ST} [given by (7-27)] over a full range of excitation frequencies, it is necessary to select the spring and damping constants as given by the parameters α and β, respectively, so that the amplitude x_0 of the primary mass is a minimum. The optimum value of α is given by

$$\alpha_{OPT} = \frac{1}{1+\mu} \tag{7-31}$$

or

$$\xi_{OPT} = \sqrt{\frac{\mu}{2(1+\mu)}} \tag{7-32}$$

Isolation of Shock[20-24]

The application of isolators to alleviate the effects of shock is difficult to discuss, partly because the term "shock" has no definite and accepted meaning. It seems to connote suddenness, either in the application of a force or in the inception of a motion. Two general types of problems are encountered in the application of isolators to reduce shock. In one type of problem, the applied force is of an impulsive nature, with the consequence that a massive member acquires additional momentum. The motion associated with this increase in momentum introduces a requirement that the isolator have adequate energy-storage capacity to arrest such motion. In a nonimpulsive application of force, the machine upon which such force acts often acquires a motion temporarily but arrests itself ultimately after experiencing displacement. The optimum isolator is one that permits such motion while preventing the transmission of excessive force. Such an isolator is not required to have capacity to store energy but only to provide a suitable support for the equipment.

Impulsive Loading. Consider the two-spring, two-mass system shown in Figure 7-36. The body m_y represents a machine supported by an isolator k_y, which in turn is supported by a foundation represented by the system m_s and k_s. The machine m_y suddenly acquires a downward velocity when acted upon by the impulsively applied force F. The effectiveness of the isolator k_y is indicated by a reduced deflection of the spring k_s [recall (7-17)].

A simplified analysis of this system may be carried out by making the following assumptions: (1) the downward velocity of the mass m_y is acquired instantaneously; (2) the isolator k_y is linear; (3) the natural frequency of the machine and isolator system $f_{n_y} = (1/2\pi)\sqrt{k_y/m_y}$ is small relative to the natural frequency of the support $f_{n_s} = (1/2\pi)\sqrt{k_s/m_s}$; and

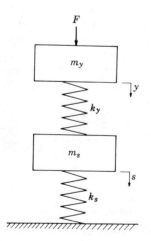

FIGURE 7-36
Schematic diagram to illustrate the effectiveness of isolator k_y in protecting foundation m_s and k_s from an impulsively applied force F acting on mounted machine m_y.

(4) the mass m_y of the machine is negligible relative to the mass m_s of the support.

When the machine is rigidly mounted to the floor ($k_y \to \infty$) the following expression gives the maximum deflection s_0 of the foundation

$$s_0 = \frac{J}{2\pi f_{n_s} m_s \sqrt{1 + \dfrac{m_y}{m_s}}} \qquad (7\text{-}33)$$

where J is the impulse of the force F in kg-sec. The influence of the mass of the machine m_y on reducing the maximum displacement is shown in Figure 7-37a.

The effectiveness of an isolator is investigated by letting k_y be relatively small in accordance with above assumptions (3) and (4). This creates two independent systems that are uncoupled because m_y is assumed much smaller than m_s. The maximum displacement y_0 of the machine is then given by

$$y_0 = \frac{J}{2\pi f_{n_y} m_y} = \frac{J}{\sqrt{k_y m_y}} \qquad (7\text{-}34)$$

When the machine is supported on isolators, assuming $m_y \ll m_s$, the maximum deflection s_0 is

$$s_0 = \frac{J}{2\pi f_{n_s} m_s} \left(\frac{f_{n_y}}{f_{n_s}} \right) \qquad (7\text{-}35)$$

Equation 7-35 is shown in Figure 7-37b. Inasmuch as the maximum deflection of the foundation is directly proportional to the natural

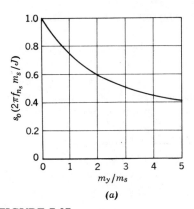

(a)

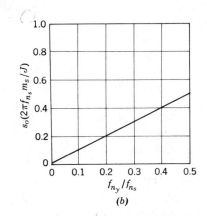

(b)

FIGURE 7-37
Effects of ways to reduce the deflection s_0 of the foundation m_s and k_s in Figure 7-36.

frequency of the isolator system, the value s_0 may be decreased by either increasing m_y or decreasing k_y. The latter may be undesirable because it increases the motion of the machine as indicated in (7-34). The addition of mass to a machine supported by isolators, that is, an increase in m_y in (7-34) and (7-35), decreases both the deflection of the support and the movement of the mounted machine. The combination of isolators with an inertia block (m_s) for the machine is an optimum solution to problems requiring energy storage within the isolators where there is an impulsive addition of energy.

Nonimpulsive Loading. If there is no change in the overall momentum of the system but only a change in the distribution of momentum within the system, it is not essential that the isolators have energy-storage capacity but rather only freedom to permit motion as dictated by the momentum transformation. A typical punch press is indicative of this type of momentum transfer.

For this type of loading, the isolators are selected so that their natural period is substantially greater than the duration of the applied force. The natural period of the isolator system must not coincide, however, with the period between the applications of the force.

Typical Isolation Materials and Configurations[25-27]

The choice of resilient material for any given application depends largely on the required deflection. Points to be considered are life, chemical stability, cost, and damping.

Damping is a result of internal friction in the resilient material. In the normal operating range, damping reduces the isolating efficiency. A certain amount of damping is desirable, however, because it reduces the vibrational motion of the machine during startup and shutdown when the speed passes through resonance.

There are several ways in which rubber mounts can be arranged as vibration isolators. These are illustrated in Figure 7-38. For rubber in compression, shown in Figure 7-38a, room must be allowed for the mount to bulge out so that a reasonable vertical flexibility is obtained. However, compression mounts may, at times, have too much horizontal flexibility. Figure 7-38b shows rubber in tension, which is not safe to use because the bond can fail. Rubber in shear, which is shown in Figure 7-38c, is widely used because it permits substantial deflections. The particular mount depicted, however, does not offer enough horizontal flexibility for most applications. The last configuration in Figure 7-38d incorporates a balance between horizontal and vertical flexibility by suitable arrangement of material to get both compression and shear in each direction.

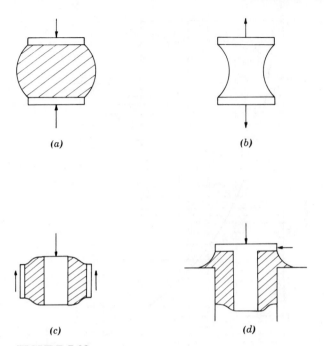

FIGURE 7-38
Various loading configurations for rubber mounts.

Cork, which has a very high internal damping and a limited flexibility, will seldom isolate primary vibrations. Cork pads are useful, however, for acoustic isolation. For very light loads, felt is often used instead of cork.

Steel springs, which have a very high flexibility and very low damping, are used to obtain a high degree of isolation at low speeds. Steel springs have practically unlimited life, and are, therefore, preferred for applications where isolator replacement must be avoided.

Figure 7-39 shows the typical displacement range and, therefore, frequency range [recall (7-21)], over which these different materials are used.

Several typical designs of commercial spring mounts are shown in Figure 7-40. Housings are needed for stability when the springs are slender. Misalignment of the two housings, caused by incorrect installation or lateral thrusts from pressure differentials across the isolated machinery, often bind the resilient snubbers, which separate the housings. Under extreme conditions there may even be metal to metal contact. This reduction in isolation efficiency can be prevented by the use of open-type spring mounts, having an outside diameter approximately equal to the working

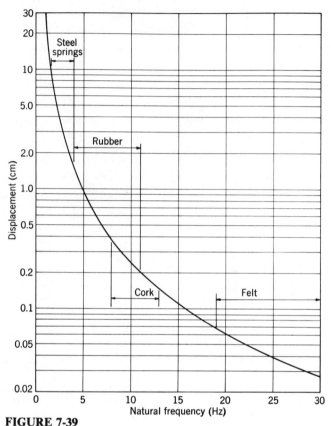

FIGURE 7-39
Displacements ranges for various isolation materials.

height for adequate stability. Large horizontal thrusts should be controlled with spring control units.

Special vibration isolators for concrete block bases can be embedded when the concrete is poured at the site into a form resting on the floor. After installation, the machine and the inertia block are raised off the floor by taking up on the leveling bolts of the isolators. (See Figure 7-40c.)

It should be noted that all the spring mounts shown in Figure 7-40 feature an acoustical pad between the spring and the foundation. These pads, which can be made of rubber, neoprene, or cork, prevent audible frequency transmission due to internal resonances within the spring elements. Such wave effects occur in rubber mounts, but to a much lesser degree.

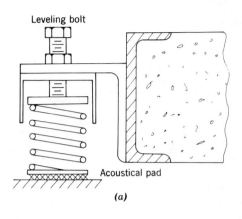

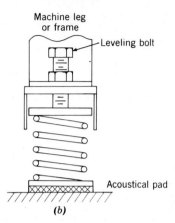

(a)

(b)

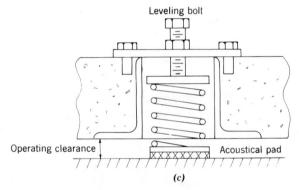

(c)

FIGURE 7-40
Typical vibration isolator configurations using steel springs.

Mounts made of knitted stainless steel wire or of plastic-enclosed glass fiber are available for special installations requiring high damping.

Example 7-4. A fan running at 700 rpm is driven by an electric motor running at 1500 rpm. The assembly is to be mounted on steel springs that deflect 13 mm. Determine the transmissibility resulting from a slight imbalance of each machine.

Solution: Since steel springs are used it is assumed that $\xi = 0$. From (7-21) (or Figure 7-39)

$$f_n = \sqrt{\frac{980.7}{2\pi}} \; \sqrt{\frac{1}{\delta_{ST}}} \; = 4.98 \sqrt{\frac{1}{\delta_{ST}}} \; = \frac{4.98}{\sqrt{1.3}} = 4.37 \text{ Hz}$$

Using (7-20) and (7-23) with $\xi = 0$ gives

$$T = \left| 1 - \left(\frac{f}{4.37} \right)^2 \right|^{-1}$$

The frequency of vibration of the fan is $700/60 = 11.67$ Hz corresponding to a transmissibility of

$$T = \left| 1 - \left(\frac{11.67}{4.37} \right)^2 \right|^{-1} = 0.163$$

This corresponds to a decrease in the transmitted force of $-20\log_{10}(0.163) = 15.8$ dB. The frequency of vibration of the motor is $1500/60 = 25$ Hz, which corresponds to a transmissibility of

$$T = \left| 1 - \left(\frac{25}{4.37} \right)^2 \right|^{-1} = 0.0315$$

or a decrease in the transmitted force of 30.0 dB. These results could have also been obtained from Figure 7-31.

Example 7-5. A machine running at 1800 rpm weighs 1000 kg. It is isolated with four steel springs placed along its base so that each mount carries the same weight. What should be the deflection of each spring to decrease the force transmission 40 dB from its original level.

Solution: Assuming $\xi = 0$, a rearrangement of (7-23), (7-21), and (7-20) yields

$$\delta_{ST} = \frac{g}{4\pi^2 f^2} \left(\frac{1}{T} - 1 \right)$$

Since $T = 10^{-40/20} = 0.01$ and $f = 1800/60 = 30$ Hz, the above relation gives

$$\delta_{ST} = \frac{980.7}{4\pi^2 (30)^2} \left(\frac{1}{0.01} - 1 \right) = 2.73 \text{ cm}$$

Example 7-6. A machine weighing 220 kg has a rotational speed of 975 rpm. The maximum amplitude of vibration at this speed is 0.36 mm. Assuming that the four vibration isolators are not fully compressed and are undamped, determine the extra weight to be added to the machine to reduce the vibration amplitude by 10 dB. It may also be assumed that the exciting force is constant and the spring constant has a value of 45 kg/cm.

Solution: The original displacement x_0 is determined from (7-19), (7-20), and (7-21) with $\xi=0$. Thus

$$x_0 = \frac{F}{k}\left|1-\left(\frac{f}{f_0}\right)^2\right|^{-1}$$

where $f_0 = (1/2\pi)\sqrt{k/m_0}$ and $m_0 = 55/g$. The new displacement amplitude x_n, is

$$x_n = \frac{F}{k}\left|1-\left(\frac{f}{f_n}\right)^2\right|^{-1}$$

where $f_n = (1/2\pi)\sqrt{k/m_n}$, $m_n = (55+W_a)/g$ and W_a is one quarter of the total weight to be attached to decrease x_0 by 10 dB. Thus

$$\frac{x_0}{x_n} = \left|1-\left(\frac{f}{f_n}\right)^2\right|\cdot\left|1-\left(\frac{f}{f_0}\right)^2\right|^{-1}$$

which can be solved for f_n^2 to yield

$$\frac{1}{f_n^2} = \frac{4\pi^2 m_n}{k} = \frac{1}{f^2}\left|1-\frac{x_0}{x_n}+\frac{x_0}{x_n}\frac{f^2}{f_0^2}\right|$$

Solving for W_a gives

$$W_a = \frac{kg}{4\pi^2 f^2}\left|1-\frac{x_0}{x_n}+\frac{x_0}{x_n}\frac{f^2}{f_0^2}\right|-55 \qquad \text{kg}$$

Since $f = 975/60 = 16.25$ Hz, $f_0 = (1/2\pi)\sqrt{(980.7)(45)/(55)} = 4.51$ Hz, and $x_0/x_n = 10^{10/20} = 3.16$, the above expression yields

$$W_a = \frac{(45)(980.7)}{4\pi^2(16.25)^2}\left|1-3.16+3.16\left(\frac{16.25}{4.51}\right)^2\right|-55$$

$$= 164.5 - 55 = 109.5 \text{ kg}$$

Therefore a total weight of $(4)(109.5) = 438.1$ kg must be added to the machine, which is twice the original weight of the machine.

7.4   TRANSMISSION LOSS AND IMPACT ISOLATION

Introduction

The sound insulation provided by internal walls and floors, and by the exterior walls and roof of a building, is an important factor in the control of noise. Transmission of noise into a room of a building can take place in several ways: (1) by air-borne sound causing the walls or floor structures to vibrate and, therefore, radiate sound; (2) by the transmission of vibration in the building structure caused by mechanical vibration or impact imparted directly to it; thus vibration may also cause a surface, such as a wall or floor, to vibrate and to radiate sound; and (3) entirely through the air; for example, by entry of sound through open windows and doors and the ducts of ventilating systems.

Air-borne sound is transmitted through a wall in the following way. Sound waves incident on one side of a wall exert a fluctuating pressue on it. As a result, the wall vibrates and radiates sound into the space on the opposite side. For most practical constructions, the heavier the partition, the smaller will be its amplitude of vibration and, therefore, the greater sound insulation it will afford. The transfer of energy from an air-borne sound wave to a much denser material in any useful wall construction is low. Only a small fraction of the energy in the incident sound wave is transformed into vibrational energy of the wall and then radiated as sound on the far side of the wall. The major portion is either reflected or absorbed at the surface on the side exposed to the sound. However, some of this energy will be transmitted along the floor to adjacent wall partitions or floors. The transmission of sound from one room to another by some path other than through the panel (wall or floor) directly in the path of the sound is called *flanking transmission*. Flanking transmission can often be the limiting factor in controlling the propagation of noise from one room to another.

When the sound incident on a wall is of the same frequency as one of the natural frequencies of the wall, the wall will resonate and vibrate at a much larger amplitude than at other frequencies. The result will be that a proportionally larger part of the incident sound will be transmitted. The insulation will be particularly low at the lowest natural frequency. If the effects of resonances are to be avoided, it is desirable to have the lowest natural frequency as low as possible. This condition can be met by partitions of large mass and small stiffness.

Coincidence Effect

An effect analogous to resonance occurs when a thin homogeneous partition, constructed of material with low damping, is set into flexural vibration by plane sound waves incident to the partition at an oblique angle. At certain frequencies the phases of the incident sound waves will "coincide" with the phase of vibration of the panel in such a way as to transmit a high fraction of the incident sound. Figure 7-41* demonstrates that at certain angles of incidence, the wavelength of the flexural wave in

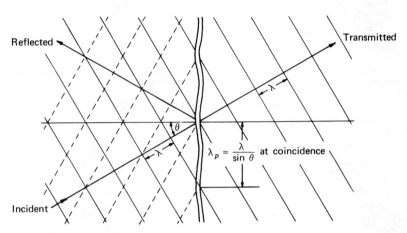

FIGURE 7-41
Coincidence effect: sound obliquely incident on a panel supporting a bending wave.
(Reprinted, by permission, from Reference 29.)

the partition is such that the pressure crests in the sound wave coincide with the crests in the flexural wave. This occurs when the wavelength in the partition is $\lambda/\sin\theta$, where λ is the wavelength of sound in air and θ is the angle of incidence ($\theta=0°$ corresponds to a normal incidence). In a uniform plate of thickness h and of very large extent the critical frequency, f_c, is

$$f_c = \frac{0.551}{h}\frac{c^2}{c_p} \quad \text{Hz} \tag{7-36}$$

where c is the speed of sound in air and $c_p{}^2 = E/[\rho(1-\nu^2)]$, E is the tensile modulus of the plate, ρ its mass density, and ν its Poisson's ratio. Equation 7-36 is plotted for some common materials in Figure 7-42. The coincidence effect can be reduced by the use of very stiff and thick walls, which greatly decreases f_c or by heavy walls with small stiffness, which greatly increases f_c.

*Figures 7-41, 7-43, and 7-44 are used with the permission of the Controller of Her Britannic Majesty's Stationery Office.

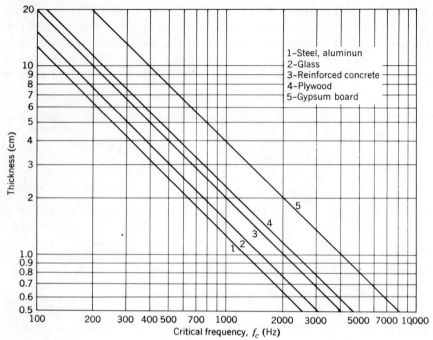

FIGURE 7-42
Critical frequency as a function of thickness for several common materials.

Sound Transmission Loss

A quantitative measure of the air-borne sound insulation of a structure is called the sound-transmission loss, TL, which is defined as the number of decibels by which sound energy randomly incident on a partition is reduced in transmission through it. Notice that the TL for partitions is defined and measured for the condition of random sound incidence, that is, in the presence of a diffuse field.

Suppose a plane wave of sound is incident on a partition. A certain fraction of the energy will be transmitted through the structure. This fraction will vary with frequency and the angle of incidence. At a given frequency, the fraction that represents the average over all angles of incidence is defined as the transmission coefficient τ. The transmission loss is related to τ by

$$TL = 10\log_{10}\frac{1}{\tau} \qquad \text{dB} \quad 0 < \tau \leqslant 1 \qquad (7\text{-}37)$$

It should be noted that when $\tau = 1$, the wall transmits 100% of the sound energy incident upon it. This corresponds to a TL of 0 dB.

Laboratory Measurement of Transmission Loss*

A specimen of the wall under test is placed in an aperture between two test rooms that are acoustically insulated from each other. Only the wall under test transmits sound from the source room to the receiving room. A diffuse sound field is created in the source room by means of loudspeakers radiating noise of a certain bandwidth (usually 1/3-octave). The sound transmitted through the test wall gives rise to a random sound field in the reverberant receiving room. The sound level in the receiving room is then determined by the sound power entering and by the total absorption of the room. The transmission coefficient and the TL of the test wall are deduced from the following formula:

$$\text{TL} = L_1 - L_2 + 10\log_{10}\frac{S}{R_T} \qquad \text{dB} \qquad (7\text{-}38)$$

where S is the area of the test wall, L_1 and L_2 are the sound-pressure levels in the source and receiving room, respectively, and R_T is the total number of sound absorptive units in the receiving room given by (6-17). Equation (7-38) is sometimes rewritten in a normalized form by adding a "correction" term that takes into account the absorption in the receiving room, R_T, relative to a standard absorption, R_0. Then the normalized level difference, D_N, is given by

$$D_N = L_1 - L_2 + 10\log_{10}\frac{R_0}{R_T} \qquad \text{dB} \qquad (7\text{-}39)$$

where $R_0 = 10 \text{ m}^2$ of absorption.

Mass Law

One of the simplest types of wall construction is a nonporous, homogeneous partition, such as brick walls, concrete walls, and solid plaster partitions. At a given frequency, the normal incidence transmission loss, $(\text{TL})_0$, for a homogeneous partition is given by

$$(\text{TL})_0 = 10\log_{10}\left[1 + \left(\frac{\pi m f}{\rho c}\right)^2\right] \qquad \text{dB} \qquad (7\text{-}40)$$

where m is the mass per unit area of the panel, ρ and c are the mass density and speed of sound of the air, respectively, and f is the frequency. Equation 7-40 is valid for $f < f_c$, typically $f < f_c/2$, where f_c is the critical

*See References 19–21 of Chapter 6.

frequency given by (7-36). When $\pi mf/(\rho c)\gg 1$, (7-40) simplifies to

$$(\text{TL})_0 = 20\log_{10}\frac{\pi mf}{\rho c} = -42.2 + 20\log_{10}fW_p \qquad \text{dB} \quad f<f_c \quad (7\text{-}41)$$

where W_p is the mass of the partition in kg/m². Equation 7-41 is called the mass law, which shows that $(\text{TL})_0$ increases 6 dB per doubling of either the mass of the partition or the frequency.

If the sound is randomly incident from all angles from 0 to 80° rather than just normal to the panel, the averge TL is given by

$$(\text{TL})_R \approx (\text{TL})_0 - 5 \qquad \text{dB} \quad f<f_c \qquad\qquad (7\text{-}42)$$

or, upon using (7-41)

$$(\text{TL})_R \approx -47.2 + 20\log_{10}fW_p \qquad \text{dB} \quad f<f_c \qquad (7\text{-}43)$$

Equation 7-43 is shown in Figure 7-43 with the $(\text{TL})_R$ of several common materials. The dip in the curves corresponds to the coincidence frequency effects.

At frequencies above the critical frequency, $(\text{TL})_R$ can be approximated by[28]

$$(\text{TL})_R \approx (\text{TL})_0 + 10\log_{10}\frac{2\eta}{\pi}\frac{f}{f_c} \qquad \text{dB} \quad f>f_c \qquad (7\text{-}44)$$

where η is the loss factor of the panel, including energy losses resulting from radiation of sound from the panel. The value of η varies between

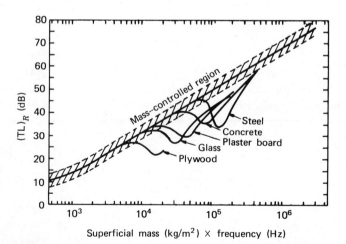

FIGURE 7-43

Relation between $(\text{TL})_R$ and the product of W_p and f showing the coincidence effect for several materials. (Reprinted, by permission, from Reference 29.)

0.005 and 0.03 depending on the material and construction techniques. Representative values for several materials can be found in Beranek.*

For the simplest types of wall constructions an empirical relationship between the mean TL (within $\pm$ 3 dB) over the frequency range 100–3150 Hz and the weight of the partition, W_p (kg/m²), is given by

$$(TL)_{mean} = 10 + 14.5 \log_{10} W_p \qquad dB \qquad (7\text{-}45)$$

Equation (7-45) is shown in Figure 7-44.

A more sophisticated analysis of single panels using the statistical energy analysis technique is given by Crocker and Price.[30]

Transmission Loss of a Composite, Single-Panel Wall

When a wall is constructed having differing transmission coefficients $\tau_1, \tau_2, \ldots,$ and corresponding areas $S_1, S_2, \ldots,$ the total sound-power transmitted through the composite wall is

$$\bar{\tau} = \frac{1}{S} \sum_i S_i \tau_i \qquad (7\text{-}46)$$

where $S = \Sigma S_i$ is the total area of the wall and $\bar{\tau}$ is now the average

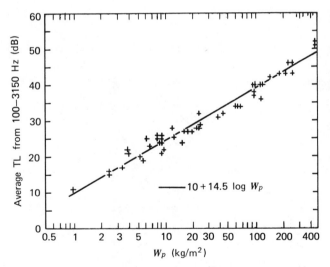

FIGURE 7-44
Relation between mean value of the TL over the frequency range 100 to 3150 Hz and W_p for single, approximately homogeneous partitions. (Reprinted, by permission, from Reference 29.)

*Reference 10 of Chapter 6, pp. 308–309.

transmission coefficient of the entire wall of area S. From (7-37) we have

$$R_c = 10\log_{10}\frac{1}{\overline{\tau}} \qquad \text{dB} \qquad (7\text{-}47)$$

where R_c is the transmission loss of the composite wall.

For a two-element wall (7-46) and (7-47) yield

$$R_c = R_1 - 10\log_{10}\left[1 - \frac{S_2}{S} + \left(\frac{S_2}{S}\right)10^{(R_1 - R_2)/10}\right] \qquad \text{dB} \qquad (7\text{-}48)$$

where

$$R_i = -10\log_{10}\tau_i \qquad \text{dB} \qquad i = 1, 2$$

It should be noted that $(S_2/S)\times 100$ is the percentage of the total area having a TL equal to R_2. Equation 7-48 is shown in Figure 7-45. Equation 7-48 gives a good estimate of the combined sound transmission loss as can be seen by comparing the measured and calculated values for a typical exterior wall, depicted in Figure 7-46. The wall in this example is of wood siding containing glazed insulating glass covering 20% of the total area.

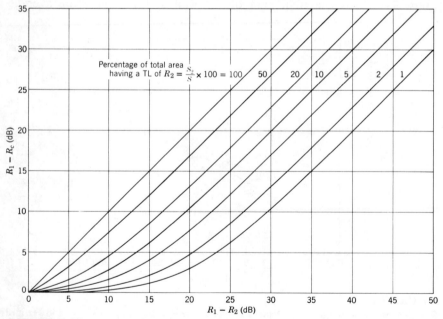

FIGURE 7-45
Transmission loss of a two-element composite barrier as a function of the relative transmission loss of the components.

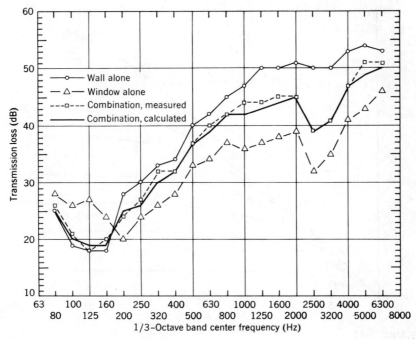

FIGURE 7-46
Comparison of measured and calculated transmission loss of a composite wall.

When the TL of one of the walls is 0 dB we have a leak (opening) in the partition. In this case $R_2 = 0$ dB and (7-48) becomes

$$R_c = R_1 - 10\log_{10}\left(1 - \frac{S_2}{S} + \frac{S_2}{S}10^{R_1/10}\right) \qquad \text{dB} \qquad (7\text{-}49)$$

where now $(S_2/S) \times 100$ is the percentage of the total area containing the opening. Equation 7-49 is shown in Figure 7-47. The significance of Figure 7-47 or (7-49) is as follows: Consider a solid wall having a TL of 47 dB. It is found that the original wall now contains openings comprising 0.1% of its total area. Since $R_1 = 47$ dB an extrapolation of Figure 7-47 [or use of (7-49)] shows the wall to now have a TL = 30 dB. Thus Figure 7-47 shows that even the smallest openings can greatly reduce the efficacy of a high-TL wall.

More sophisticated attempts to predict the sound transmission loss through holes in panels can be found in the literature.[31-33]

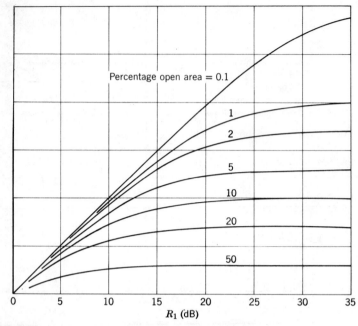

FIGURE 7-47
The effect of a hole of a given percentage of the total area on a partition originally having a TL of R_1 dB.

Transmission Loss of Double Panels

Consider two panels of mass per unit area m_1 and m_2 separated by an air gap of width d. The TL for the double panels is given by the following three expressions:[28]

$$\frac{\rho c}{\pi M} \ll f < f_0$$

$$(\text{TL})_D \approx -42.4 + 20\log_{10} Wf \qquad \text{dB} \qquad (7\text{-}50a)$$

$$f_0 < f < \frac{c}{2\pi d}$$

$$(\text{TL})_D \approx (\text{TL})_1 + (\text{TL})_2 + 20\log_{10} 2kd \qquad \text{dB} \qquad (7\text{-}50b)$$

$$\frac{c}{2\pi d} < f$$

$$(\text{TL})_D \approx (\text{TL})_1 + (\text{TL})_2 + 6 \qquad \text{dB} \qquad (7\text{-}50c)$$

where $M = m_1 + m_2$, W is the total weight of both panels (kg/m^2), $k = 2\pi f/c$ is the wave number, and $(\text{TL})_1$ and $(\text{TL})_2$ are average transmission losses of each panel given by (7-43) with W_p replaced by W_1 and W_2, respectively. The quantity f_0 is the natural frequency of the two panels coupled by the air cushion in the cavity and is given by

$$f_0 = \frac{c}{2\pi} \left[\frac{\rho}{d} \left(\frac{1}{m_1} + \frac{1}{m_2} \right) \right]^{1/2} \text{Hz}$$

From (7-50a) it is evident that the TL of the double panel in the indicated frequency region increases 6 dB per doubling of either frequency or mass of each panel. In the frequency range presented for (7-50b) the TL increases 18 dB per doubling of frequency. In the range suggested for (7-50c) the TL increases 12 dB per doubling of either the frequency or the mass of each panel. Equations 7-50a, b, and c do not hold, of course, at frequencies in the vicinity of and greater than the critical frequencies of either panel. Since these results assumed that each panel behaves independently of the other, the $(\text{TL})_j$ $(j = 1, 2)$ are given by (7-44) in this frequency region.

Several more sophisticated attempts at theoretically determining the TL of double panels can be found in the literature.[34-36]

Single-Figure Ratings: STC and IIC

It is often convenient to provide a single-figure rating that can be used for comparing partitions for general building design purposes. The ratings are designed to correlate with subjective impressions of the sound insulation provided against the sounds of speech, radio, television, music, and similar sources of noise in offices and dwellings. Excluded from the single-figure ratings given below are applications involving noise spectra that differ markedly from those stated above. Thus noises such as those produced by most machinery, certain industrial processes, bowling alleys, power transformers, and so forth, would be excluded. A particular omission would be the exterior walls of buildings, for which noise problems are most likely to involve motor vehicles and aircraft, or both. In all such cases, it is best to use the detailed sound transmission loss values, in conjunction with the actual spectra of the intrusive noises.

STC—Sound Transmission Class

To determine the sound transmission class (STC) of a test specimen, its sound transmission losses in a series of 16 1/3-octave bands are compared to those of the reference contour (the STC contour) shown in Figure 7-48. The center frequency of the first 1/3-octave band is 125 Hz and that of the 16th is 4000 Hz. The STC is usually determined graphically. The transmission loss for the test specimen is plotted as a function of the center frequencies of the 1/3-octave bands on the same scale as the STC contour shown in Figure 7-48. The STC is then determined by comparison with a transparent overlay of Figure 7-48 on which the STC contour is drawn. The STC contour is shifted vertically, relative to the test curve, until some of the measured TL values for the test specimen fall below those of the STC contour (the solid line) and the following conditions are fulfilled:

1. The sum of the deficiencies (that is, the deviations below the contour) shall not be greater than 32 dB.
2. The maximum deficiency at a single test point shall not exceed 8 dB [the broken (dashed) line beneath the STC contour].

When the contour is adjusted to the highest value (in integral dB) that meets the above requirements, the sound transmission class for the specimen is the TL value corresponding to the intersection of the contour and the 500-Hz ordinate. The TL values of the ordinate are those given by (7-38). It should also be noted that the greater the STC value, the more sound insulation it provides.

Figure 7-49 shows two examples of the determination of the STC values for two different walls.

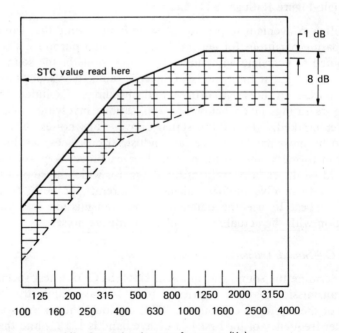

1/3−Octave band center frequency (Hz)

FIGURE 7-48
Overlay from which the STC is determined graphically.

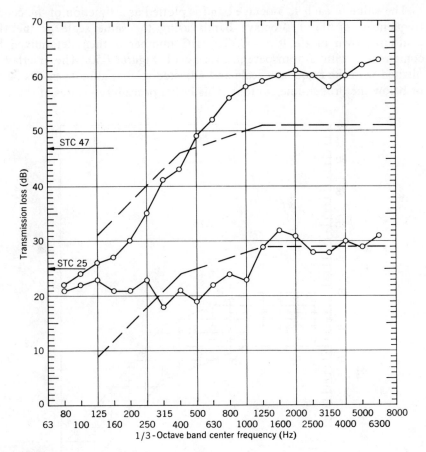

FIGURE 7-49
Determination of the STC value for two different walls from measured TL data.

IIC—Impact Isolation Class

The impact isolation class (IIC) is a single-number rating that provides a means for comparing the acoustical performance of floor-ceiling assemblies when excited by impacts produced by a standard tapping machine. The resulting impact-sound-transmission levels (ISTL) are analyzed in 1/3-octave bands and normalized according to the relation

$$L_{\text{ISTL}} = L_2 + 10\log_{10}\frac{R_T}{R_0} \qquad \text{dB} \qquad (7\text{-}51)$$

where L_2 is the sound-pressure level in the 1/3-octave band and R_T and R_0 are the same as those appearing in (7-39). The 1/3-octave bands are the space-averaged sound-pressure levels from 100 to 3150 Hz.

The value in each 1/3-octave band is plotted as a function of the center frequencies of the 1/3-octave bands using the same scale as the IIC contour shown in Figure 7-50. The IIC number is then determined by comparison with a transparent overlay of Figure 7-50. The overlay is aligned with the frequency scale and adjusted so that all data points lie on or below the broken-line contour. This initial procedure insures at the start

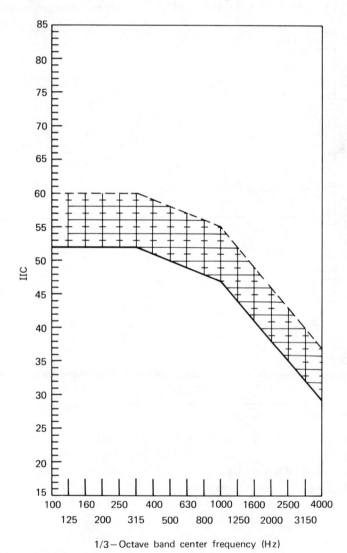

1/3−Octave band center frequency (Hz)

FIGURE 7-50
Overlay from which IIC is determined graphically.

that single deviations are less than or equal to 8 dB. The deviations above the solid-line contour are then summed. The total must not exceed 32 dB; if greater, the overlay is adjusted upward until the total is equal to 32 dB or the nearest integer not exceeding 32 dB. The IIC value, *read from the overlay*, is that value which lies over (corresponds to) the impact-sound-pressure level of 60 dB on the graph scale.

INR—Impact Noise Rating

The impact noise rating is another single-figure rating method required by the FHA. The procedure is exactly the same as that used to obtain the IIC except that the shape of the overlay contour is slightly different and the assignment of the single-figure rating is arrived at by the following procedure: After the INR contour has been adjusted to satisfy the deviation requirements, the intersection of the horizontal portion of the contour with the vertical scale is determined. This value, in dB, is subtracted from 66 dB. Thus if the horizontal portion of the INR contour falls below 66 dB, the INR has a plus value, and if it falls above the 66 dB, the INR has a minus value. A floor-ceiling having a rating of 0 INR has been designated by FHA from field experience as marginally acceptable for impact isolation under actual occupancy. Additionally, positive INR's indicate better performance and minus values poorer.

Recommended STC and IIC Values for Dwellings

The U. S. Department of Housing and Urban Development[37] has established criteria for walls and floors of multi-family housing. These criteria, which take into account consideration of geographic locations, economic conditions, and functions of partitions and floors, are presented in terms of three grades. Grade I is applicable primarily in suburban and peripheral urban residential areas, which might be considered as the "quiet" locations for background noise. Nighttime exterior A-weighted noise levels may be 35–40 dB. In addition, the insulation criteria of this grade are applicable in certain special cases, such as dwelling units above the eighth floor in high-rise buildings and "luxury" buildings, regardless of location. Grade II is the most important category and is applicable primarily in residential urban and suburban areas considered to have the "average" noise environment. Nighttime exterior A-weighted noise levels might be 40–45 dB. Grade III criteria should be considered as minimal recommendations and are applicable in some urban areas that are generally considered "noisy" locations, with nighttime exterior A-weighted noise levels of 45 dB or higher.

The criteria for wall systems are given in Table 7-1 and those for floor systems in Table 7-2.

TABLE 7-1
HUD Recommendations for STC Values of Walls in Multi-family Housing

Partition function between dwellings			STC		
Apt. A	to	Apt. B	Grade I	Grade II	Grade III
Bedroom			55	52	48
Living room			57	54	50
Kitchen		Bedroom	58	55	52
Bathroom			59	56	52
Corridor			55	52	48
Living room			55	52	48
Kitchen		Living room	55	52	48
Bathroom			57	54	50
Corridor			55	52	48
Kitchen			52	50	46
Bathroom		Kitchen	55	52	48
Corridor			55	52	48
Bathroom			52	50	46
Corridor		Bathroom	50	48	46

TABLE 7-2
HUD Recommendations for STC and IIC Values of Floors in Multi-family Housing

Partition function between dwellings			Sound transmission class (STC)			Impact isolation class (IIC)		
			Grade			Grade		
Apt. A	above	Apt. B	I	II	III	I	II	III
Bedroom			55	52	48	55	52	48
Living room			57	54	50	60	57	53
Kitchen		Bedroom	58	55	52	65	62	58
Family room			60	56	52	65	62	58
Corridor			55	52	48	65	62	58
Bedroom			57	54	50	55	52	48
Living room			55	52	48	55	52	48
Kitchen		Living room	55	52	48	60	57	53
Family room			58	54	52	62	60	56
Corridor			55	52	48	60	57	53

TABLE 7-2 (Continued)

Partition function between dwellings			Sound transmission class (STC) Grade			Impact isolation class (IIC) Grade		
Apt. A	above	Apt. B	I	II	III	I	II	III
Bedroom			58	55	52	52	50	46
Living room			55	52	48	55	52	48
Kitchen		Kitchen	52	50	46	55	52	48
Bathroom			55	52	48	55	52	48
Family room			55	52	48	60	58	54
Corridor			50	48	46	55	52	48
Bedroom			60	56	52	50	48	46
Living room		Family room	58	54	52	52	50	48
Kitchen			55	52	48	55	52	50
Bathroom		Bathroom	52	50	48	52	50	48
Corridor		Corridor	50	48	46	50	48	46

TABLE 7-3
Typical STC Values for Various Types of Glass and Plastics

Glass Type	STC
Single strength	19
Double strength	21
3.2 mm rolled	23
4.8 mm sheet or rolled	25
5.6 mm sheet or rolled	25
6.4 mm sheet, plate, or float	26
9.8 mm plate	27
9.5 mm plate or sheet	29
12.7 mm plate or sheet	31
6.4 mm laminated, 0.381 mm plastic inner layer	31
25.4 mm insulating with 12.7 mm air space	32
15.9 mm solid plate	34
7.1 mm laminated, heavy plastic inner layer	36
19.1 mm solid plate	36
22.2 mm solid plate	36
25.4 mm solid plate	37
12.7 mm laminated, heavy plastic inner layer	40
19.1 mm laminated, heavy plastic inner layer	43

The large amount of varied and detailed data available giving both the transmission loss as a function of the 1/3-octave frequency bands and the corresponding STC/IIC values for many of the floor and wall systems in current use can be obtained from references 38 through 61. Table 7-3 gives some typical STC values for glass partitions. Figures 7-51 and 7-52 give the configuration and corresponding typical STC values for common floor and wall constructions.

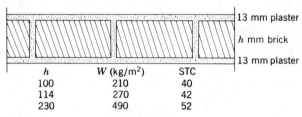

h	W (kg/m²)	STC
100	210	40
114	270	42
230	490	52

FIGURE 7-51
Typical STC values of various common wall constructions. (1) Brick. (From *Environmental Acoustics* by L. L. Doelle. Copyright 1972 by McGraw-Hill Book Company. Used with permission of McGraw-Hill Book Company.)

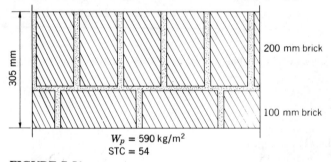

$W_p = 590$ kg/m²
STC = 54

FIGURE 7-51
(2) Brick—305 mm.

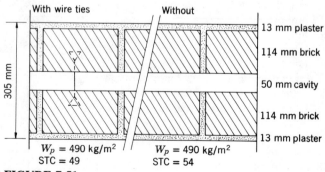

$W_p = 490$ kg/m² $W_p = 490$ kg/m²
STC = 49   STC = 54

FIGURE 7-51
(3) Cavity brick.

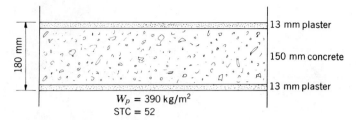

FIGURE 7-51
(4) Concrete—150 mm.

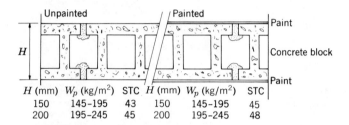

H (mm)	W_p (kg/m²)	STC	H (mm)	W_p (kg/m²)	STC
150	145–195	43	150	145–195	45
200	195–245	45	200	195–245	48

FIGURE 7-51
(5) Hollow dense concrete block.

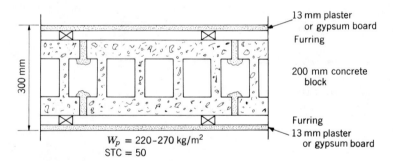

FIGURE 7-51
(6) Hollow dense concrete block—200 mm.

FIGURE 7-51
(7) Gypsum wallboard.

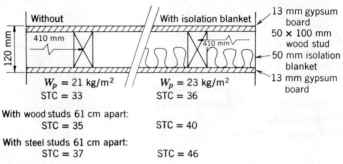

$W_p = 21$ kg/m² $W_p = 23$ kg/m²
STC = 33   STC = 36

With wood studs 61 cm apart:
     STC = 35   STC = 40

With steel studs 61 cm apart:
     STC = 37   STC = 46

FIGURE 7-51
(8) Wood stud partitions—50 × 100 mm.

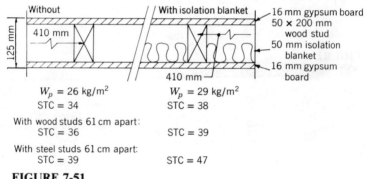

$W_p = 26$ kg/m² $W_p = 29$ kg/m²
STC = 34   STC = 38

With wood studs 61 cm apart:
     STC = 36   STC = 39

With steel studs 61 cm apart:
     STC = 39   STC = 47

FIGURE 7-51
(9) Wood stud—50 × 100 mm.

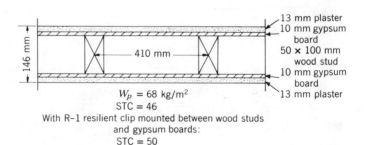

$W_p = 68$ kg/m²
STC = 46
With R-1 resilient clip mounted between wood studs
and gypsum boards:
STC = 50

FIGURE 7-51
(10) Wood stud—50 × 100 mm.

278

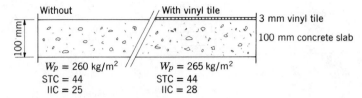

FIGURE 7-52

Typical STC and IIC values of various common floor constructions. (1) Concrete slab—100 mm. (From *Environmental Acoustics* by L. L. Doelle. Copyright 1972 by McGraw-Hill Book Company. Used with permission of McGraw-Hill Book Company.)

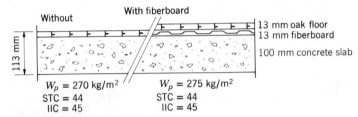

FIGURE 7-52
(2) Concrete slab—100 mm.

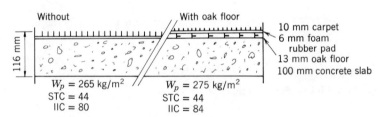

FIGURE 7-52
(3) Concrete slab—100 mm.

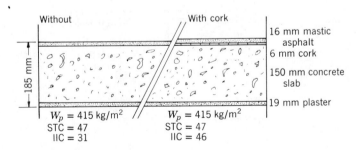

FIGURE 7-52
(4) Concrete slab—150 mm.

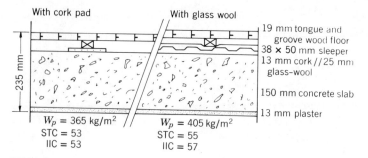

With cork pad   With glass wool

19 mm tongue and groove wood floor
38 × 50 mm sleeper
13 mm cork // 25 mm glass-wool
150 mm concrete slab
13 mm plaster

235 mm

$W_p = 365 \text{ kg/m}^2$
STC = 53
IIC = 53

$W_p = 405 \text{ kg/m}^2$
STC = 55
IIC = 57

FIGURE 7-52
(5) Concrete slab—150 mm.

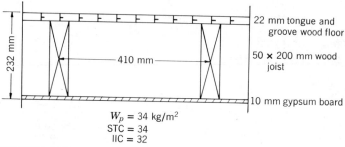

232 mm

410 mm

22 mm tongue and groove wood floor

50 × 200 mm wood joist

10 mm gypsum board

$W_p = 34 \text{ kg/m}^2$
STC = 34
IIC = 32

FIGURE 7-52
(6) Wood joist—50 × 200 mm.

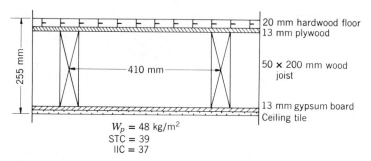

255 mm

410 mm

20 mm hardwood floor
13 mm plywood

50 × 200 mm wood joist

13 mm gypsum board
Ceiling tile

$W_p = 48 \text{ kg/m}^2$
STC = 39
IIC = 37

FIGURE 7-52
(7) Wood joist—50 × 200 mm.

280

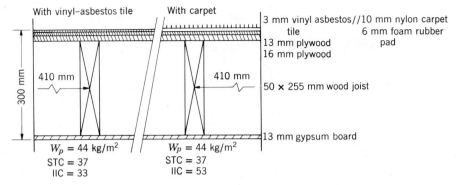

With vinyl-asbestos tile   With carpet

3 mm vinyl asbestos // 10 mm nylon carpet
tile   6 mm foam rubber
13 mm plywood   pad
16 mm plywood

410 mm   410 mm

50 × 255 mm wood joist

300 mm

13 mm gypsum board

W_p = 44 kg/m^2 W_p = 44 kg/m^2
STC = 37   STC = 37
IIC = 33   IIC = 53

FIGURE 7-52
(8) Wood joist—50 × 255 mm.

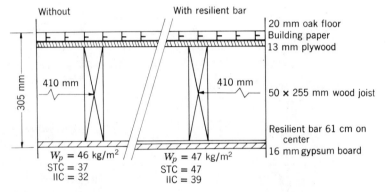

Without   With resilient bar

20 mm oak floor
Building paper
13 mm plywood

410 mm   410 mm

305 mm

50 × 255 mm wood joist

Resilient bar 61 cm on
center
16 mm gypsum board

W_p = 46 kg/m^2 W_p = 47 kg/m^2
STC = 37   STC = 47
IIC = 32   IIC = 39

FIGURE 7-52
(9) Wood joist—50 × 255 mm.

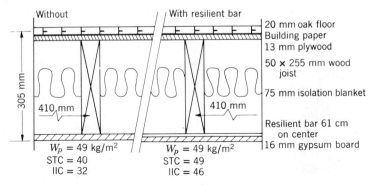

Without   With resilient bar

20 mm oak floor
Building paper
13 mm plywood

50 × 255 mm wood
joist

75 mm isolation blanket

410 mm   410 mm

305 mm

Resilient bar 61 cm
on center
16 mm gypsum board

W_p = 49 kg/m^2 W_p = 49 kg/m^2
STC = 40   STC = 49
IIC = 32   IIC = 46

FIGURE 7-52
(10) Wood joist—50 × 255 mm.

281

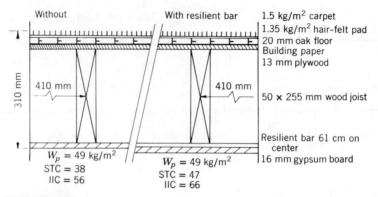

FIGURE 7-52
(11) Wood joist—50 × 255 mm.

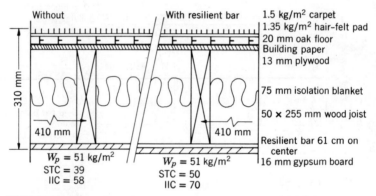

FIGURE 7-52
(12) Wood joist—50 × 255 mm.

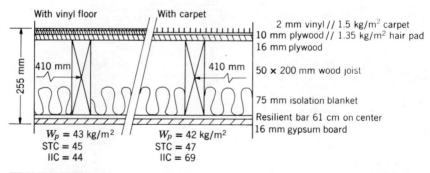

FIGURE 7-52
(13) Wood joist—50 × 200 mm.

282

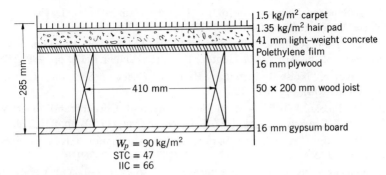

FIGURE 7-52
(14) Wood joist—50 × 200 mm.

Relationship Between TL and L_{ISTL}

The relationship between the normalized sound transmission loss, TL, given by (7-38) and the normalized impact sound transmission level, L_{ISLT}, given by (7-51) has been shown to be[62]

$$L_{ISTL} + TL = 43 + 30\log_{10}f_B \quad dB \qquad (7-52)$$

where f_B is the center frequency of the octave or 1/3-octave band under consideration. Equation 7-52 is valid for very hard floors (e.g., concrete) of moderate thickness (> 10 cm). If this floor is "floated," that is, if it rests on a resilient mount (7-52) becomes

$$L_{ISTL} + TL = 43 + 30\log_{10}f_B - 10\log_{10}\left(1 + \frac{f_B^4}{f_1^4}\right) \quad dB \qquad (7-53)$$

where f_1 is the natural frequency of the floor-slab-resilient-mount system. These results have been independently experimentally verified.[63]

REFERENCES

1. D. D. Davis, Jr., "Acoustic Filters and Mufflers," Chapter 21 in *Handbook of Noise Control*, C. M. Harris, Ed., McGraw-Hill, New York (1957).

2. A. V. Sreenath and M. L. Munjal, "Evaluation of Noise Attenuation Due to Exhaust Mufflers," *J. Sound Vib.*, Vol. 12, No. 1 (1970), pp. 1–19.

3. R. J. Alfredson and P.O.A.L. Davis, "Performance of Exhaust Silencer Components," *J. Sound Vib.*, Vol. 15, No. 2 (1971), pp. 175–196.

4. T. Wu, "Control of Diesel Engine Exhaust Noist," Paper No. 700701, Society of Automotive Engineers (1970).

5. R. J. Alfredson and P.O.A.L. Davis, "The Radiation of Sound from an Engine Exhaust," *J. Sound Vib.*, Vol 13, No. 4 (1970), pp. 389–408.

6. G. J. Sanders, "Noise Control for Industrial Air Moving Devices," Inter-Noise 72 Proceedings (October 1972), pp. 165–170.

7. "Power Plant Acoustics," Technical Manual TM 5-805-9, Headquarters, Department of the Army, Washington, D. C. (December 1968), pp. 280–289.

8. U. J. Kurze, "Noise Reduction by Barriers," *J. Acoust. Soc. Am.*, Vol 55, No. 3 (March 1974), pp. 504–518.

9. A. D. Pierce, "Diffraction of Sound Around Corners and Over Wide Barriers," *J. Acoust. Soc. Am.*, Vol. 55, No. 5 (May 1974), pp. 941–954.

10. G. F. Butler, "A Note on Improving the Attenuation Given by a Noise Barrier," *J. Sound Vib.*, Vol. 32, No. 3 (1974), pp. 367–369.

11. Z. Maekawa, "Noise Reduction by Screens," *Appl. Acoust.*, Vol. 1 (July 1968), pp. 157–173.

12. U. J. Kurze and G. S. Anderson, "Sound Attenuation of Barriers," *Appl. Acoust.*, Vol. 4 (January 1971), pp. 35–53.

13. M. Koyasu and M. Yamashita, "Scale Model Experiments in Noise Reduction by Acoustic Barrier of a Straight Line Source," *Appl. Acoust.*, Vol. 6 (1973), pp. 233–242.

14. W. E. Scholes, A. C. Salvidge, and J. W. Sargent, "Barriers and Traffic Noise Peaks," *Appl. Acoust.*, Vol. 5 (1972), pp. 205–222.

15. W. E. Scholes and J. W. Sargent, "Designing Against Noise from Road Traffic," *Appl. Acoust.*, Vol. 4 (1971), pp. 203–233.

16. "Noise Barrier Design and Example Abatement Measures," U. S. Department of Transportation, Federal Highway Administration, National Highway Institute, Washington, D. C. (April 1974).

17. D. I. Cook and D. F. Van Haverbeke, "Trees and Shrubs for Noise Abatement," Research Bulletin 246, The Forest Service, U. S. Department of Agriculture in Cooperation with University of Nebraska College of Agriculture, Lincoln, Neb. (July 1971).

18. D. I. Cook and D. F. Van Haverbeke, "Tree-Covered Land Forms for Noise Control," Research Bulletin 263, The Forest Service, U. S. Department of Agriculture, in cooperation with Institute of Agriculture and Natural Resources, University of Nebraska, Lincoln, Neb. (July 1974).

19. "Fundamentals and Abatement of Highway Traffic Noise," Report No. FHWA-HH1-HEV-73-7976-1, U. S. Department of Transportation, Federal Highway Administration, Washington, D. C. (June 1973), Chapter 5.

20. C. M. Harris and C. E. Crede, Ed., *Shock and Vibration Handbook*, Vols. I and II, McGraw Hill, New York (1961), Chap. 3, 6, 30, and 34.

21. C. E. Crede, *Shock and Vibration Concepts in Engineering Design*, Prentice-Hall, Englewood Cliffs, N. J. (1965), Chapter 4.

22. Chapters 12 and 13 of Reference 1.

23. J. E. Ruzicka, "Passive Shock Isolation, Part I," *Sound Vib.* (August 1970), pp. 14–24.

24. J. E. Ruzicka, "Passive Shock Isolation, Part II," *Sound Vib.* (September 1970), pp. 10–22.

25. C. M. Salerno, "How to Select Vibration Isolators for Use as Machinery Mounts," *Sound Vib.* (August 1973), pp. 22–29.

26. R. M. Huchheiser, "How to Select Vibration Isolators for DEM Machinery Equipment," *Sound Vib.* (August 1974), pp. 14–23.

27. R. Adair, "The Design and Application of Pneumatic Vibration Isolators," *Sound Vib.* (August 1974), pp. 24–27.

28. B. H. Sharp, "A Study of Techniques to Increase the Sound Insulation of Building Elements," Wyle Laboratories, El Segundo, Calif. (June 1973) (NTIS No. PB-222 829).

29. E. N. Bazley, "The Airborne Souna insulation of Partitions," National Physical Laboratory, London, Her Majesty's Stationary Office (1966).

30. M. J. Crocker and A. J. Price, "Sound Transmission Using Statistical Energy Analysis," *J. Sound Vib.*, Vol. 9, No. 3 (1969), pp. 469–486.

31. G. Westerberg, "On the Sealing of Circular Holes in a Thick Wall for the Purpose of Sound Insulation," *Appl. Acoust.*, Vol 4 (1971), pp. 115–129.

32. M. C. Gomperts and T. Kihlman, "The Sound Transmission Loss of Circular and Slit-Shaped Apertures in Walls," *Acustica*, Vol. 18 (1967), pp. 144–150.

33. A. Sauter, Jr. and W. W. Soroka, "Sound Transmission through Rectangular Slots of Finite Depth Between Reverberent Rooms," *J. Acoust. Soc. Am.*, Vol. 47, No. 1 (1970), pp. 5–11.

34. A. J. Price and M. J. Crocker, "Sound Transmission Through Double Panels Using Statistical Energy Analysis," *J. Acoust. Soc. Am.*, Vol. 47, No. 3 (1970), pp. 683–693.

35. R. J. Donato, "Sound Transmission Through a Double-Leaf Wall," *J. Acoust. Soc. Am.*, Vol. 57, No. 3 (1972), pp. 807–815.

36. M. J. Crocker, M. C. Bhattacharya, and A. J. Price, "Sound and Vibration Transmission through Panels and Tie Beams Using Statistical Energy Analysis," *ASME J. Eng. Ind.* (August 1971), pp. 775–782.

37. U. S. Department of Housing and Urban Development, Publication FT-TS 24 (1968).

38. R. D. Berendt, G. E. Winzer, and C. B. Burroughs, "A Guide to Airborne, Impact and Structural Borne Noise Control in Multifamily Dwellings," U. S. Department of Housing and Urban Development, Washington, D. C. (September 1967).

39. T. D. Northwood, "Transmission Loss of Plasterboard Walls," Building Research Note No. 66, National Research Council, Ottawa, Canada (October 1968, revised July 1970).

40. D. R. Prestemon, "Least-Cost Wall and Floor Construction for Limiting Transmission of Noise," Report No. 55, Agricultural and Home Economics Experiment Station, Iowa State University of Science and Technology, Ames, Iowa (February 1968).

41. "Solutions to Noise Control Problems," Owens-Corning Fiberglas Corp., Home Building Products Division, Toledo, Ohio (1969).

42. R. D. Berendt and G. E. Winzer, "Sound Insulation of Wall, Floor and Door

Constructions," National Bureau of Standards Monograph 71 (November 1964).

43. W. E. Purcell, "Compendium of Materials for Noise Control," Final Technical Report, IITRI Project 56285, Engineering Mechanics Division, IIT Research Institute, Chicago, Ill. (April 1974). (Prepared for Department of Health, Education, and Welfare, National Institute for Occupational Safety and Health, Cincinnati, Ohio.)

44. H. J. Sabine, "Sound and Thermal Transmission and Air Infiltration of Residential Exterior Walls, Doors, and Windows," Final Report to National Bureau of Standards, Center for Building Technology, Washington, D. C. (November 1972).

45. "Acoustical Manual: Apartment and Home Construction," National Association of Home Builders, NAHB Research Foundation, Inc., Rockville, Md. (June 1971).

46. J. B. Grantham and T. B. Heebink, "Sound Attenuation Provided by Several Wood-Framed Floor-Ceiling Assemblies with Troweled Floor Toppings," *J. Acoust. Soc. Am.*, Vol. 52, No. 2 (1973), pp. 353–360.

47. K. A. Mulholland, "Sound Insulation Measurements on a Series of Double Plasterboard Panels with Various Infills," *Appl. Acoust.*, Vol. 4 (1971), pp. 1–12.

48. W. Loney, "Effect of Cavity Absorption and Multiple Layers of Wallboard on the Sound-Transmission Loss of Steel-Stud Partitions," *J. Acoust. Soc. Am.*, Vol. 53, No. 6 (1973), pp. 1530–1534.

49. R. D. Ford and G. Kerry, "The Sound Insulation of Partially Open Double Glazing," *Appl. Acoust.*, Vol. 6 (1973), pp. 57–72.

50. J. A. Marsh, "The Airborne Sound Insulation of Glass," *Appl. Acoust.*, Vol. 4 (1971), pp. 175–191.

51. W. Loney, "Effect of Cavity Absorption on the Sound Transmission Loss of Steel-Stud Gypsum Wallboard Partitions," *J. Acoust. Soc. Am.*, Vol. 49 No. 2 (1971), pp. 385–390.

52. R. D. Ford, P. Lord, and P. C. Williams, "The Influence of Absorbent Linings on the Transmission Loss of Double-Leaf Partitions," *J. Sound Vib.*, Vol. 5, No. 1 (1967), pp. 22–28.

53. J. B. Grantham, "Airborne Noise Control in Lightweight Floor/Ceiling Systems," *Sound Vib.* (June 1971), pp. 12–16.

54. P. B. Ostergaard, R. L. Cardinell, and L. S. Goodfriend, "Transmission Loss of Leaded Building Materials," *J. Acoust. Soc. Am.*, Vol. 35, No. 6 (June 1963), pp. 837–843.

55. T. B. Heebink, "Effectiveness of Sound Absorptive Materials in Dry Walls," *Sound Vib.* (May 1970), pp. 16–18.

56. P. W. Parkin, H. J. Purkis, and W. E. Scholes, "Field Measurements of Sound Insulation Between Dwellings," Her Majesty's Stationary Office, London (1960) (National Building Studies Research Paper No. 33).

57. E. Ellwood, "The Anatomy of a Wall," *Sound Vib.* (June 1972), pp. 14–18.

58. R. D. Ford, D. C. Hothersall, and A. C. C. Warnock, "The Impact Insulation Assessment of Covered Concrete Floors," *J. Sound Vib.*, Vol. 33, No. 1 (1974), pp. 103–115.

59. E. T. Weston, M. A. Burgess, and J. A. Whitlock, "Airborne Sound Transmission Through Elements of Buildings," Technical Study No. 48, Department of Housing and Construction, Experimental Building Station, Australian Government Publishing Service, Canberra, Australia (1973).

60. "Architectural Acoustical Materials," Annual Bulletin, Acoustical and Board Products Association, Park Ridge, Ill.

61. "Fire Resistance," Design data compiled by the Gypsum Association, Chicago, Ill. (1971–1972).

62. I. L. Ver, "Relation between the Normalized Impact Sound Level and Sound Transmission Loss," *J. Acoust. Soc. Am.*, Vol. 50, No. 6 (1971), pp. 1414–1417.

63. S. Wolf and N. J. Mason, "Flanking Transmission of Floating Floor Systems," Noise-Con 73 Proceedings (October 1973), pp. 535–539.

BIBLIOGRAPHY

NOISE CONTROL

L. H. Bell, *Fundamentals of Industrial Noise Control*, 2nd ed., Harmony Publications, Trumbull, Conn. (1974).

L. L. Beranek, Ed., *Noise and Vibration Control*, McGraw-Hill, New York (1971).

B. F. Day, R. D. Ford, and P. Lord, *Building Acoustics*, Elsevier, (1969).

G. M. Diehl, *Machinery Acoustics*, Wiley, New York (1973).

C. Duerden, *Noise Abatement*, New York Philosophical Library (1971).

C. M. Harris, Ed., *Handbook of Noise Control*, McGraw-Hill, New York (1957).

P. Jensen and G. Sweitzer, *How You Can Soundproof Your Home*, Lexington Publishing, Lexington, Mass. (1974).

A. J. King, *The Measurement and Suppression of Noise*, Chapman Hall, London (1965).

R. H. Lyon, *Lectures in Transportation Noise*, Grozier Publishing, Cambridge, Mass. (1973).

H. J. Purkis, *Building Physics: Acoustics*, Pergamon Press, New York (1966).

M. Rettinger, *Acoustic Design and Noise Control*, 2nd ed., Chemical Publishing, New York, (1973).

A. Thumann and R. K. Miller, *Secrets of Noise Control*, Fairmont Press, Atlanta, Georgia (1974).

L. F. Yerges, *Sound, Noise and Vibration Control*, Van Nostrand Reinhold, New York (1969).

Handbook of Noise and Vibration Control, 2nd ed., Trade and Technical Press, Morden, Surrey, England, (1974).

Industrial Noise Manual, 2nd ed., American Industrial Hygiene Association, Detroit, Mich., (1966).

ARCHITECTURAL ACOUSTICS

L. L. Beranek, *Music, Acoustics & Architecture*, Wiley, New York (1962).

L. L. Doelle, *Environmental Acoustics*, McGraw-Hill, New York (1972).

H. Kuttruff, *Room Acoustics*, Halsted, New York (1974).

A. B. Lawrence, *Architectural Acoustics*, Elsevier, New York (1970).

ACOUSTICS—GENERAL

L. Cremer and M. Heckl, *Structure-Borne Sound*, Translated from the German and revised by E. E. Ungar, Springer-Verlag, New York (1973).

R. D. Ford, *Introduction to Acoustics*, Elsevier, New York (1970).

L. E. Kinsler and A. R. Frey, *Fundamentals of Acoustics*, 2nd ed., Wiley, New York (1962).

E. Meyer and E. G. Neumann, *Physical and Applied Acoustics*, Academic Press, New York (1972).

P. M. Morse and K. U. Ingard, *Theoretical Acoustics*, McGraw-Hill, New York (1968).

J. W. S. Rayleigh, *The Theory of Sound*, Vol II, 2nd. ed., Dover Publications, New York (originally published 1896).

S. N. Rschevkin, *The Theory of Sound*, Pergamon Press, New York (1962).

W. C. Sabine, *Collected Papers on Acoustics*, Dover Publications, New York (1964).

E. Skudrzyk, *The Foundations of Acoustics*, Springer-Verlag, New York (1971).

R. W. B. Stephens and A. E. Bate, *Acoustics and Vibrational Physics*, Edward Arnold, London (1966).

INSTRUMENTATION

J. T. Broch, *Acoustic Noise Measurements*, 2nd ed., Brüel & Kjaer, Naerum, Denmark (1973).

D. N. Keast, *Measurements in Mechanical Dynamics*, McGraw-Hill, New York (1967).

E. B. Magrab and D. S. Blomquist, *The Measurement of Time-Varying Phenomena*, Wiley-Interscience, New York (1971).

A. P. C. Peterson and E. E. Gross, Jr., *Handbook of Noise Measurement*, 7th ed., General Radio Company, Concord, Mass. (1972).

Acoustics Handbook, Application Note 100, Hewlett-Packard Company, Palo Alto, California (November 1968).

CONFERENCE PROCEEDINGS

"Noise and Vibration Control Engineering," M. J. Crocker, Ed., Proceedings of the Purdue Noise Control Conference, Purdue University, Lafayette, Indiana (July 14–16, 1971).

"Inter-Noise 72 Proceedings," M. J. Crocker, Ed., International Conference on Noise Control Engineering, Washington, D. C. (October 4–6, 1972).

"Inter-Noise 73 Proceedings," O. J. Pedersen, Ed., International Conference on Noise Control Engineering, Technical University, Lyngby, Denmark (August 22–24, 1973).

"Noisexpo Proceedings," National Noise and Vibration Control Conference, Chicago, Illinois (September 11–13, 1973).

"Noise-Con 73 Proceedings," D. R. Tree, Ed., National Conference on Noise Control Engineering, Washington, D. C. (October 15–17, 1973).

"Inter-Noise 74 Proceedings," J. C. Snowdon, Ed., International Conference on Noise Control Engineering, Washington, D. C. (September 30–October 2, 1974).

APPENDIX A

Conversion Factors*

$$1 \text{ in.} = 0.0254 \text{ m} = 2.54 \text{ cm} = 25.4 \text{ mm}$$
$$1 \text{ in.}^2 = 6.4516 \times 10^{-4} \text{ m}^2 = 6.4516 \text{ cm}^2$$
$$1 \text{ in.}^3 = 1.639 \times 10^{-5} \text{ m}^3 = 16.39 \text{ cm}^3$$
$$1 \text{ ft} = 0.3048 \text{ m} = 30.48 \text{ cm}$$
$$1 \text{ ft}^2 = 0.0929 \text{ m}^2 = 929 \text{ cm}^2$$
$$1 \text{ ft}^3 = 0.0283 \text{ m}^3$$
$$1 \text{ ft/min} = 5.08 \times 10^{-3} \text{ m/sec}$$
$$1 \text{ ft}^3/\text{min} = 4.7195 \times 10^{-4} \text{ m}^3/\text{sec}$$
$$1 \text{ mph} = 1.609 \text{ km/hr} = 0.447 \text{ m/sec}$$
$$1 \text{ lb} = 0.4536 \text{ kg}$$
$$1 \text{ lb/in.} = 17.86 \text{ kg/m}$$
$$1 \text{ lb/in.}^2 = 703 \text{ kg/m}^2$$
$$1 \text{ lb/in.}^3 = 2.768 \times 10^4 \text{ kg/m}^3 = 27.68 \text{ g/cm}^3$$
$$1 \text{ lb/ft} = 1.488 \text{ kg/m}$$
$$1 \text{ lb/ft}^2 = 4.882 \text{ kg/m}^2$$
$$1 \text{ lb/ft}^3 = 16.02 \text{ kg/m}^3$$
$$\text{acceleration of gravity} = 9.807 \text{ m/sec}^2$$
$$°F = (9/5)°C + 32$$

*lb = pound weight.

INDEX